Preliminary Design of High-Rise Buildings in Non-Seismic Regions

Niall MacAlevey

Preface

High-rise buildings continue to be built in increasing numbers throughout the world. However, they quickly become impractically expensive if structural needs are not considered early. Similarly attempting to build an irrational design can waste a lot of money. It has been said that 50% of the cost of the structural frame costs are affected by preliminary design, whereas detailed design (e.g. refinement of reinforcement in a reinforced concrete structure) affects a small percentage only. This book discusses what should be considered at the preliminary design stage of a high-rise. The information can also be used in the checking of designs produced by computer. The focus is on European design practice. The discussion includes the following: the origin of lateral loads; vortex shedding, lateral load resisting systems; effect of wind load on tall buildings; construction tolerances; shear wall layouts; basic acceleration limits; the importance of the P-delta phenomenon. Gravity systems are no different to those of low-rise buildings so little time is spent discussing these. Although tall buildings are commonly constructed using steel or composite construction, this book focuses on concrete buildings as they are especially challenging to design. A case study is presented to illustrate the importance of preliminary design of high-rises. The aim is to produce a brief book appropriate for final year students and practioners.

To Owen and Neil

CONTENTS

Preliminary Design of High-Rise Buildings in Non-Seismic Regions

CHAPTER 1: INTRODUCTION

1.1 IMPROVING PRELIMINARY DESIGN SKILLS

Two reasons why these skills should be improved:

Planning a tall building. Bigger changes in the architectural scheme are possible at the preliminary design stage. Changes are more difficult to accommodate later.

Checking the detailed design of a tall building prepared using software, e.g., ETABS: preparing the "checking model" (a simplified model used to estimate the behaviour of the detailed model).

1.2 DEFINITION

Buildings that are sufficiently tall that lateral loading influences their design,

Note that those buildings that are significantly taller than their neighbours, e.g., a 30-storey building surrounded by 20-storey buildings, are considered "tall", but the same building surrounded by 30-storey buildings may not be.

1.3 Brief History of "High-Rise" Buildings

There are some old examples:

Rome (circa 100AD): Timber-framed buildings constructed for mass housing were around 20 m high (typically 5-7 storeys).

China (circa 600AD): Masonry Pagoda (Xi'an) 62 m high (equivalent to 18 storeys).

But the modern high-rise really began when the materials became available (initially steel and later high strength concrete) and elevators were invented.

Home Insurance Building, Chicago (1885): Skeleton of cast/wrought iron and steel but heavy masonry cladding and including elevators (10-storeys).

Masonic Building, Chicago (1892): Skeleton of steel, vertical steel trusses as bracing, elevators and curtain wall (100 m tall, "the world's first 'tallest' building").

Ingalls Office Building, Cincinnati, Ohio (1904): First all-reinforced concrete "tall" building (64 m). (Virtually all tall buildings built in US from 1910 to 1940 were steel framed).

1.4 "World's Tallest"

Here are some of the world's tallest buildings:

- Woolworth Building, New York, 1913 (222 m);
- Empire State Building, New York, 1931 (381 m);
- World Trade Centre, New York, 1972 (415 m);
- Sears/Willis Tower, Chicago, 1974 (442 m);
- Petronas Towers, Kuala Lumpur, 1999 (452 m);
- Taipei 101, Taipei, 2005 (508 m);
- World Finance Centre, Shanghai, 2008 (492 m);
- Burj Dubai, Dubai, 2010 (828 m).

1.5 How are high-rises different from low-rises?

Well clearly the loads are bigger. But there are several fundamental differences in we must be aware of:

- High-rise design is usually governed by stiffness (deflection and accelerations) not strength.
- We must try and minimize holding-down (i.e., tension on foundations): gravity load should be

channeled into the same members that resist lateral load so they experience net compression.

- Dynamic loading: High-rise buildings usually have a low natural frequency (typically less than 1 Hz). This means lateral loads often have dynamic effects. If the duration of the load (t_d) divided by the natural period of the structure (T) is greater than about 5 then the load can be considered "static". Otherwise it is a "dynamic" load. Thus a gust lasting 3 seconds is a static load for a low-rise building of period 0.3 seconds ($t_d/T = 10$), while the same load is a dynamic load on a tall building of period 9 seconds ($t_d/T = 0.33$).

1.6 Preliminary Design of High Rises

Preliminary design affects up to 50% of cost of frame construction. Minimizing reinforcement in a reinforced concrete frame affects only a few percent of the cost. (Fling 1987). It is particularly important for high-rises because of the cost of remedial action.

1.7 Premium for Height

The following figure (Figure 1.1) is adapted from (Stafford-Smith, 1991). From about 30 stories onwards there is a noticeable increase in material required to ensure satisfactory behavior.

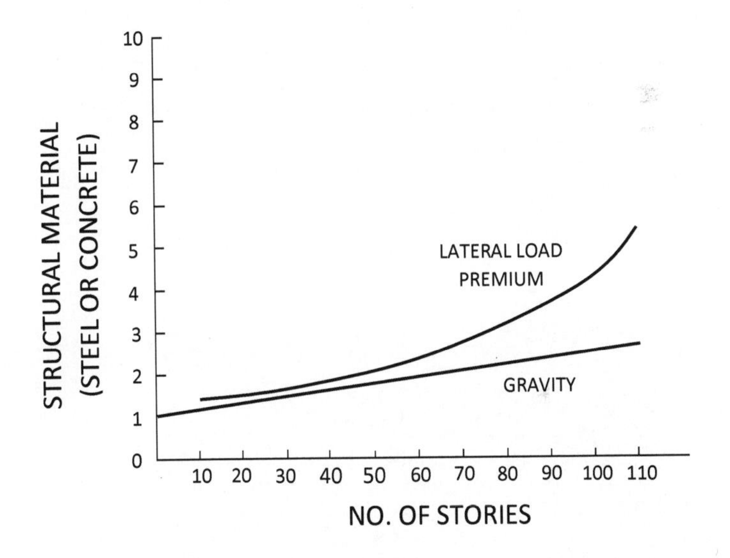

Figure 1.1: Graph showing influence of lateral load on building material required.

1.8 TALLER IMPLIES MORE SLENDER

Typically the economic slenderness of a building (i.e., Slenderness Ratio, SR = height/depth) is less than about 5-7 but buildings of SR equal to 15 have been constructed (Grossman 1990).

There is a psychological need for "outside awareness". So people "need" to be (i.e., feel more comfortable) within about 18 m of a window (Millais, 1997). Thus if the plan is fixed, higher means more slender too. (This means the largest plan dimension for a square office tower is likely to be about 60 m on plan (i.e. 18+say25+18)).

1.9 SOURCES OF LATERAL LOADS ON TALL BUILDINGS

- **Wind loads;**
- **Geometric imperfections in frame;** consisting of a) Effect on bracing system; and b) Effect on floor "diaphragm" (i.e., floor plate);
- **Earthquake loads** (not considered further).

1.9.1 Wind loads

In the past buildings had heavy cladding and partitions built tight to the structural frame. Modern buildings (i.e.

from around 1960 onwards) are generally much lighter and less stiff, so are more prone to the effects of wind. In addition, high-rise buildings are also sensitive to **vortex shedding**, **wake buffeting**, and cause **wind scoop effects.**

1.9.1.1 *Vortex Shedding*

At low wind speeds vortices are shed symmetrically from each side of the object in the airflow. At higher wind speeds (more than about 27 m/s) vortices are shed alternately. Each time one of these vortices detaches a force is exerted perpendicular to the wind (a crosswind force). Thus at these higher velocities a sideways periodic force is applied to the obstruction. The flow of wind past a building is as shown in (Figure 1.2).

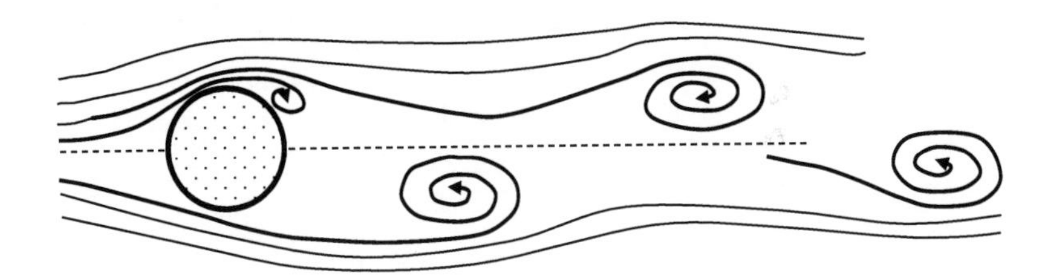

Figure 1.2: Plan of air flow past a bluff body: vortices shed alternately from each side.

The vortex shedding frequency (i.e. frequency of applied sideways load) depends on the velocity of wind and the plan-shape of building. Each shape has a characteristic number describing its behavior in the wind, called its Strouhal number. The relationship between the variables governing this complex phenomenon is surprisingly simple: $f = (v.S_t)/D$

where v is the velocity of the wind, S_t is the Strouhal number for the plan shape, and D is the building "dimension". The following table (Table 1.1) shows the effect of several building plans.

Table 1.1: Dynamic Behaviour of Various Shapes

Building Plan	**Strouhal No.**	**Vortex Shedding Behaviour**	**Crosswind Movement**
Circular	0.20	Poor	Large
Square	0.11	Good	Small
Triangle	0.16	Moderate	Medium

e.g. suppose v = 30 m/s; if the building is cylindrical of D = 30 m => S_t = 0.2,; Thus from formula, shedding frequency f = 0.2 Hz. Thus building will pulsate as if stiffness = 0 if the natural frequency of the building is about 0.2 and damping low (Taranath 1998). In other words, at a **critical wind velocity** (of 30 m/s in this case) the vortex shedding frequency is likely to coincide with the natural frequency of the building.

1.9.1.2 *Wake Buffeting (from neighbouring high-rises)*

This refers to the wind blowing off neighbouring structures. Depending on their location, the wind can be faster and more turbulent. Once the separation between buildings exceeds about 6*D* the wake buffeting effect, becomes insignificant.

1.9.1.3 *Scoop effect (on pedestrians and neighbouring low rises)*

When the wind blows against the face of a building, some of that wind is diverted down the face. This means low level winds are faster near tall buildings. This is shown in Figure 1.3 below.

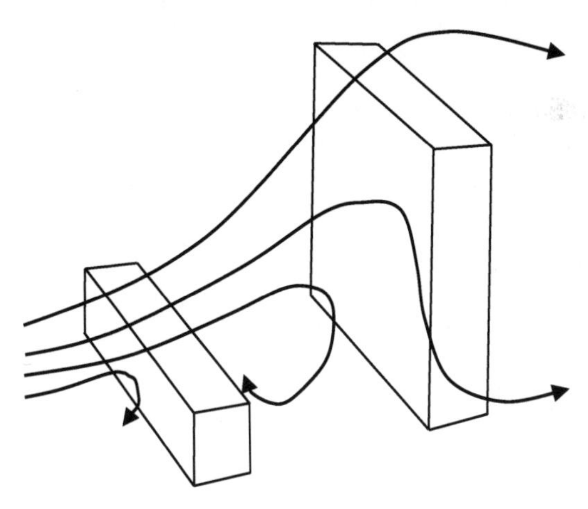

Figure 1.3: Low level winds are faster near tall buildings

1.9.1.4 *Values of wind loads*

For preliminary design the values of pressure quoted in Codes can be used especially as the results of any wind tunnel study will probably not be available. EN 1991-1-4 (EC1) gives pressures and a method for calculating accelerations (for buildings < 200 m tall).

Wind loads are less for "non-flexible" buildings (i.e., those where the slenderness = height/breadth is less than about 5) as there is little dynamic effect.

Wind gusts in Singapore/Malaysia are associated with thunderstorms. The typical value of wind speed is 32-35 m/s (3-second gust with return period 50 years).

1.9.1.5 *When is a Wind Tunnel test needed?*

A wind tunnel test is needed if **any** of these apply:

- Building 120 m or above;
- Unusual or irregular shaped building;
- Complex surroundings (e.g. risk of wake buffeting);
- Building height exceeds five times least plan dimension;

- Building's natural period, $T > 2$ seconds, (i.e. natural frequency < 0.5 Hz).
- Vortex shedding frequency similar to any natural frequency.

Numerous wind tunnel studies have shown that smoothening of corners can lead to better overall behavior. Other variations that are known to help with the wind are tapering the elevation/introduction of setbacks, slots to "spill" the wind, and spoilers to discourage the formation of vortices. However, in most cases the plan and elevation are already fixed so the calculation at preliminary design stage is of the value of the critical velocity, v_{crit}.

1.9.2 Geometric imperfections in frame

Columns are never constructed perfectly vertically: consider the consequences of this out-of-plumb (typically 1/400). Eurocode 2 requires the resulting horizontal loads to be considered in addition to wind. These imperfections affect (1) the design of the bracing system and (2) the floor diaphragm.

1) Effect on bracing system: Eurocode 2 requires consideration of the case of the building out-of-plum by α. This is shown in the figure below

(Figure 1.4). At each level the floor must equilibrate difference in vertical forces. This results in horizontal forces of approximately the same magnitude applied at each level:

$$H_i = \alpha (N_b - N_a)$$

2) Effect on floor diaphragm: Eurocode 2 requires the case of members constructed $\alpha/2$ out-of-plumb to be considered. The floor plate must equilibrate difference in vertical forces. It affects one level only.

$$H_i = \alpha (N_b - N_a)$$

where N_b and N_a are the floor loads at levels b and a respectively.

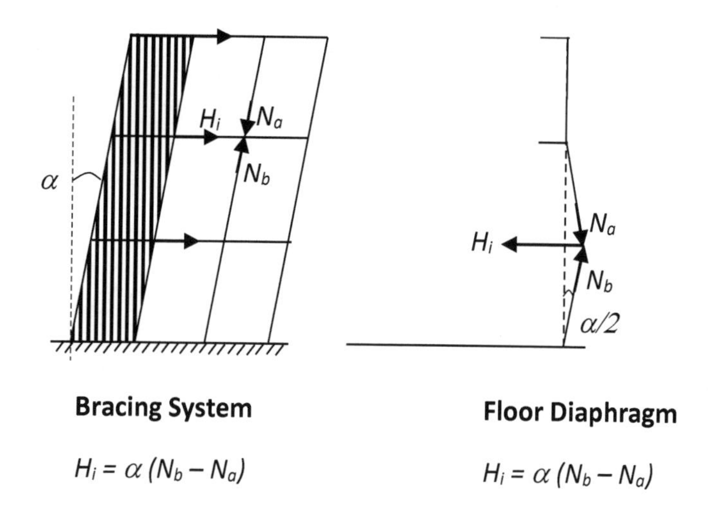

Figure 1.4: Horizontal loads from Eurocode 2

1.10 Estimates of Natural Frequency

Natural Frequency (i.e. frequency of free vibration):

$n = 46/h$ Hertz

where h is the height of the building in metres.

Period $T = 1/n$ (i.e. time taken to complete one complete cycle of free movement).

Typically for a tall building n : 0.1-0.2 Hz, i.e. T: 5-10 sec; while for a low rise building n : 1-2 Hertz, i.e. T: 0.5-1 sec.

1.11 Resistance to Lateral Loads

It is usual to assume that all lateral load is resisted by the structural frame, i.e. to ignore any resistance due to the cladding and partitions. Long-term resistance to lateral load of masonry cladding or partitions should **not** be relied upon for the following reasons:

- Shrinkage of blockwork partitions (especially if unconnected to frame),
- Creep and shrinkage of concrete frame should be allowed to occur,
- Moisture expansion of brickwork,
- Brittle failure of masonry panel means sudden transfer of load to frame,
- Partitions may be removed by tenant.

A detail ensuring isolation of partitions is shown below (Figure 1.5):

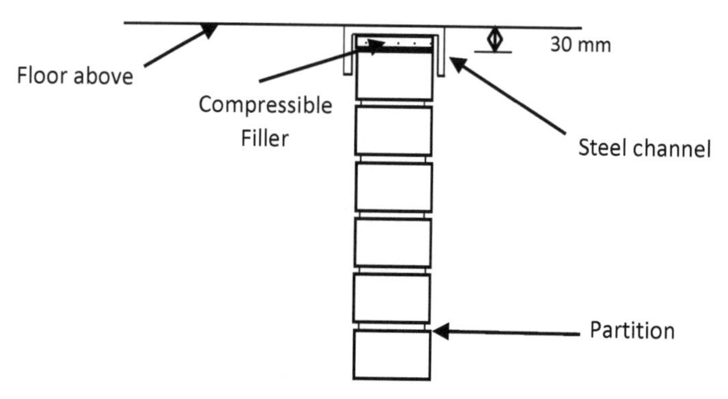

Figure 1.5: Suggested detail to restrain partition

1.12 Basement & Foundations

1.12.1 Basement:

Basements are useful structurally as they reduce the **net** pressure applied to the soil: i.e. the increase in pressure on soil is equal to the weight of building minus the weight of soil excavated. Thus they can reduce the differential (and absolute) settlement. It is very important to avoid significant differential settlement of a high-rise (see later). Thus aim to use a basement to balance a significant portion of weight to be added.

Often the lateral reaction can be taken on basement walls or vertical piles and raking piles usually can be avoided.

1.12.2 Foundations:

A raft supported by piles or using caissons is usual. Many tall buildings in Singapore use caissons up to 6m in diameter, e.g., UOB, OUB, DBS, Republic Plaza. Occasionally the raft itself is sufficient, e.g. International Plaza.

Avoid founding on materials of different stiffness, e.g. one side of high-rise on friction piles and other on rock, as the risk of differential settlement high.

Be careful of the overall weight of concrete buildings in particular. If no rock under the site, frequently the capacity of the pile group under the core is limited by **settlement**. If one has both of these characteristics, then foundation problems are likely at design stage.

1.13 Residential vs. Office Buildings

1.13.1 Hotel/apartment building:

The following are the likely characteristics of a Residential building:

- The slab is a minimum thickness of 150 mm-200 mm for sound insulation;
- A plain soffit is required for easy finishing;
- A false ceiling is required in corridors only (as the services run along the corridor ceiling).
- There is a rectangular floor plan (perhaps to ensure view) and small floor plate.
- There is a small core.
- Better daylight (natural lighting).
- More expensive to cool and clad (as not min. surface).
- Better cross-ventilation.
- Often slimmer.
- People more sensitive to accelerations.
- Regular partition layout.
- The storey height is typically around 3 m.
- The frame is usually concrete.

Typical Residential building plan: (5 units shown)

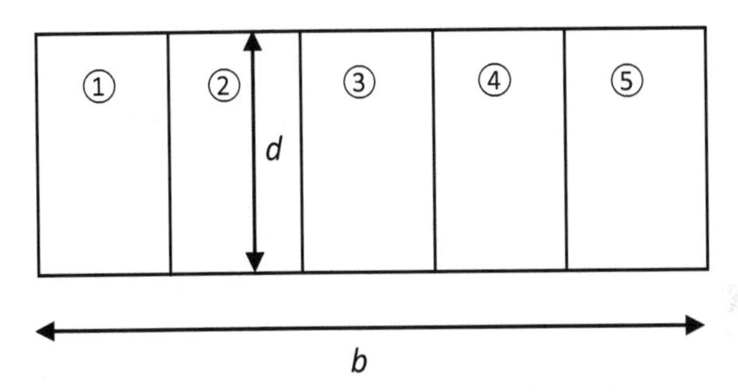

e.g., 30-storey condominium, H = 81 m. If shear walls extend full depth of building, Slenderness Ratio = H/d = 81/15 = 5.4

1.13.2 Office:

The following are the likely characteristics of an Office building:

- A false ceiling is required throughout (>1 m for services).
- The storey height is typically around 4-4.5 m.
- The frame can be concrete or steel.
- Typically square floor plan and large core.
- Large floor plates (tenants need large open plan office).
- Cheaper to cool (regular shape).

- Cheaper for cladding (min. surface area).
- More expensive to light (fluorescent lights) as floor plan deep.
- Often significantly higher.

Typical Office building plan:

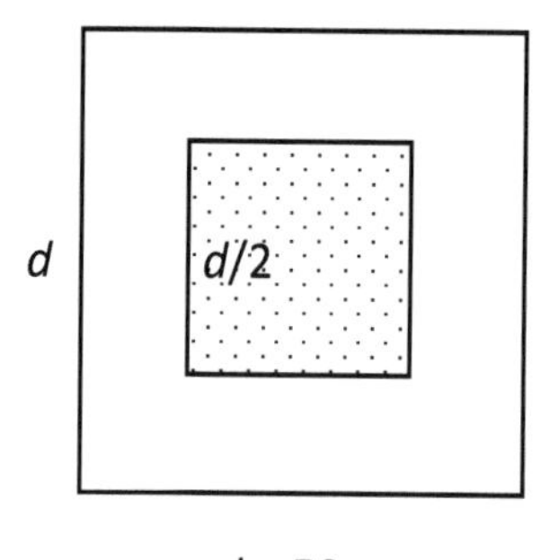

d = 50 m

H = 250 m

Calculation of slenderness should take size of lateral load resisting system into consideration

e.g.,

- core-mast system (see later): Slenderness ratio (SR) = H/d = 250/50 = 5;
- core-only system: SR = $2H/d$ = 10

Chapter 2: Lateral Load Resisting Systems

2.1 Review of Commonly used Lateral load resisting systems

- Moment-resisting frames/rigid frames;
- Braced frames;
- Shear walls;
- Framed or braced tubes;
- Tube-in-Tube Structures;
- Core Interactive Structures;
- Bundled-Tube Structures;
- Space structures;

2.1.1 Moment-resisting frames/rigid frames

Columns are connected rigidly to beams to form a moment resisting frame. These connections are relatively expensive in steel as site welding is often involved. The bending of columns and beams resist lateral load. Shorter spans mean internal frames are difficult to avoid; these conflict with leasing requirements.

A commonly used system in low- to medium-rise buildings. The maximum building height is about 25 storeys. The main structural disadvantage is that horizontal movements due to lateral load are relatively large. The deflected shape is shown in Figure 2.1.

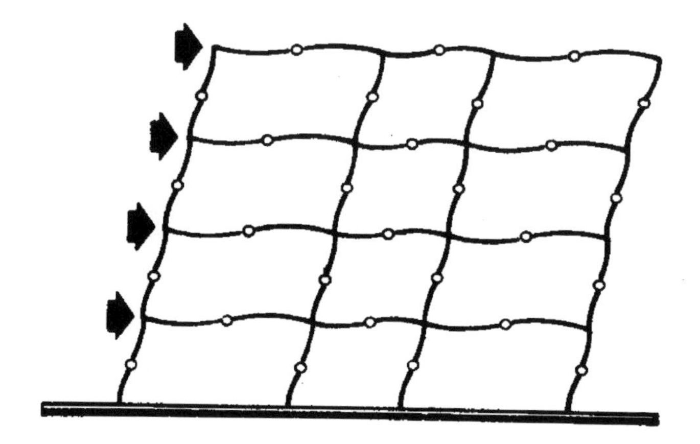

Figure 2.1: Deflected shape of moment resisting frame ("unbraced frame")

2.1.2 Braced Frames (steel)

The characteristic of the system is diagonals linking columns to form a vertical truss. It is stiffer than a moment-resisting frame. The system is common in low- to medium- rise buildings and is often used with other systems in high-rise buildings. Connections may be simple or rigid. Bracing is usually only possible in the core area, unless it is covered by cladding or expressed.

2.1.3 Shear Walls/Structural Walls

One of simplest and most popular systems for concrete buildings. The first shear-wall structures appeared in 1950s. Shear walls behave as vertical cantilevers. The floors act as rigid as beams ('diaphragms') spanning between the walls. The walls can "coupled" i.e., joined together. Frame interaction can be included at the detailed design stage (Figure 2.3). Shear walls are usually considered essential as bracing, if floors are flat slabs. The action of the shear wall/floor diaphragm is shown in Figure 2.2.

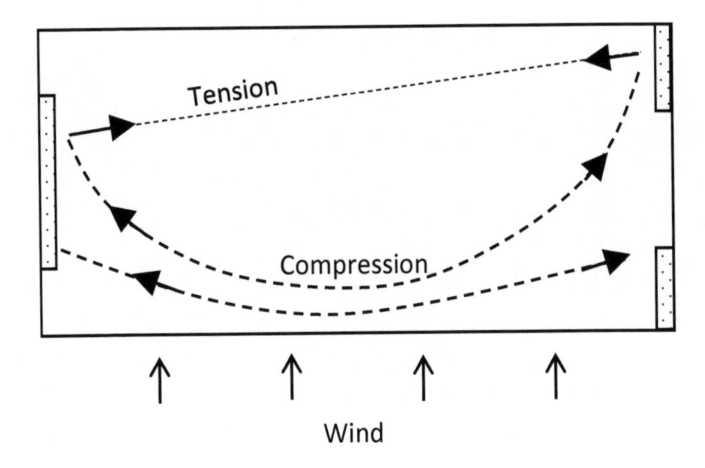

Figure 2.2: Load path for wind load from exterior to structural walls via diaphragm action

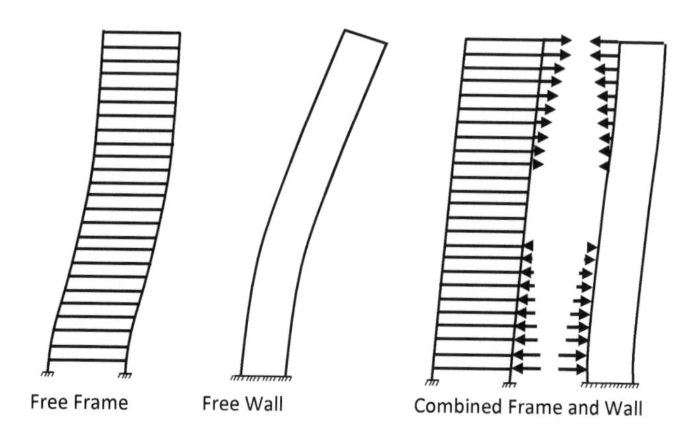

Figure 2.3: Diagram showing the effect of interaction of walls and frame

2.1.4 Framed or Braced tubes

Tube principle: lateral load resisting mechanism is placed as far away from centre of building as possible, thereby maximizing its second moment of area.

Thus all elements resisting the lateral load are placed along the perimeter. External columns are closely spaced and moment-connected to spandrel beams which are relatively deep. Thus internal columns do not carry bending and so are smaller. Perimeter columns are spaced up to 4.5 m if concrete and 6 m if steel. There is no internal bracing. In concrete buildings, flat slab construction can be used.

Framed tube structures are appropriate only for buildings up to about 60 storeys because of shear-lag. (Figure 2.4). To understand shear-lag in this situation, consider use of the "engineer's beam theory" formula, "stress = moment/section modulus". This would suggest the stress is constant along the perimeter. In reality, the flexibility of the perimeter structure ensures a non-uniform distribution.

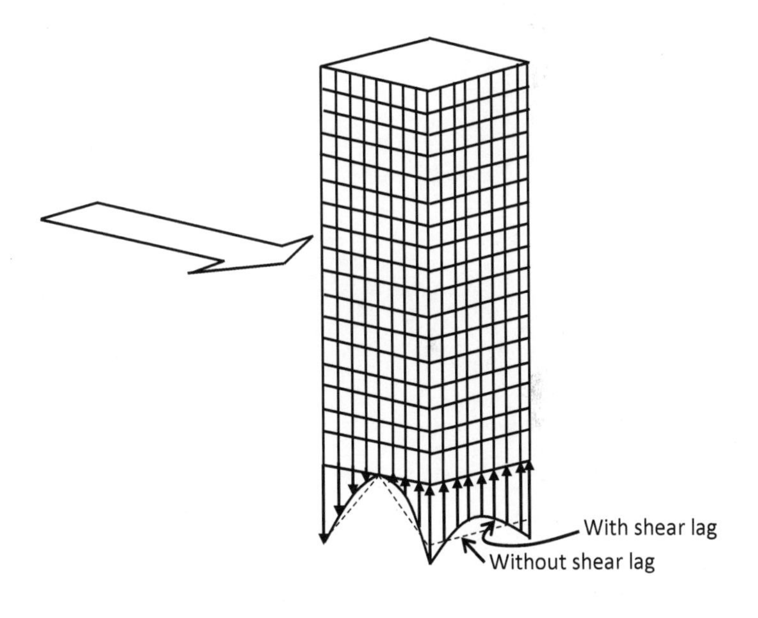

Figure 2.4: The shear-lag phenomenon in a framed tube

Braced tubes are a refinement of framed tubes which attempt to overcome the shear-lag problem. Belt trusses wrap around the perimeter at mechanical or refuge floors, to even out the column axial loads and so reduce the shear-lag (Figure 2.5). Braced tubes are appropriate for very tall structures. Alternatively, cross bracing may be used, e.g., Hancock Centre, Chicago.

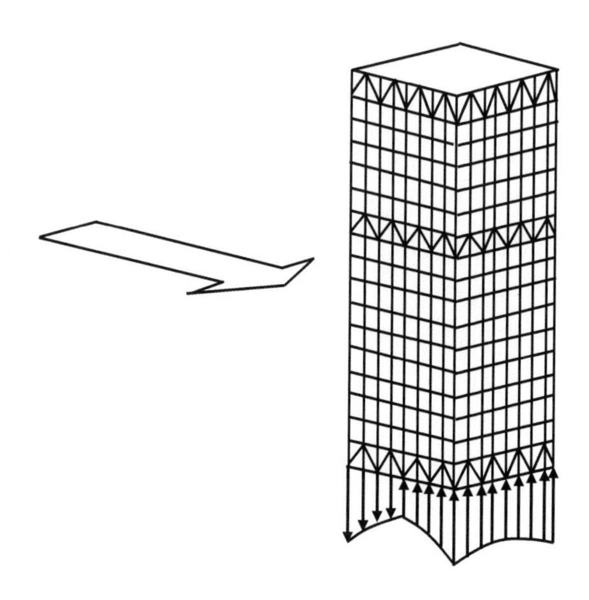

Figure 2.5: Use of belt trusses around perimeter to reduce shear-lag (three shown)

Construction of a steel tube is made easier by the use of prefabricated units. Figure 2.6 shows a two storey prefabricated "tree". The tube may also be constructed using concrete.

Nevertheless, many architects dislike the "wall-like" perimeter of tube. This has led to the development of (f).

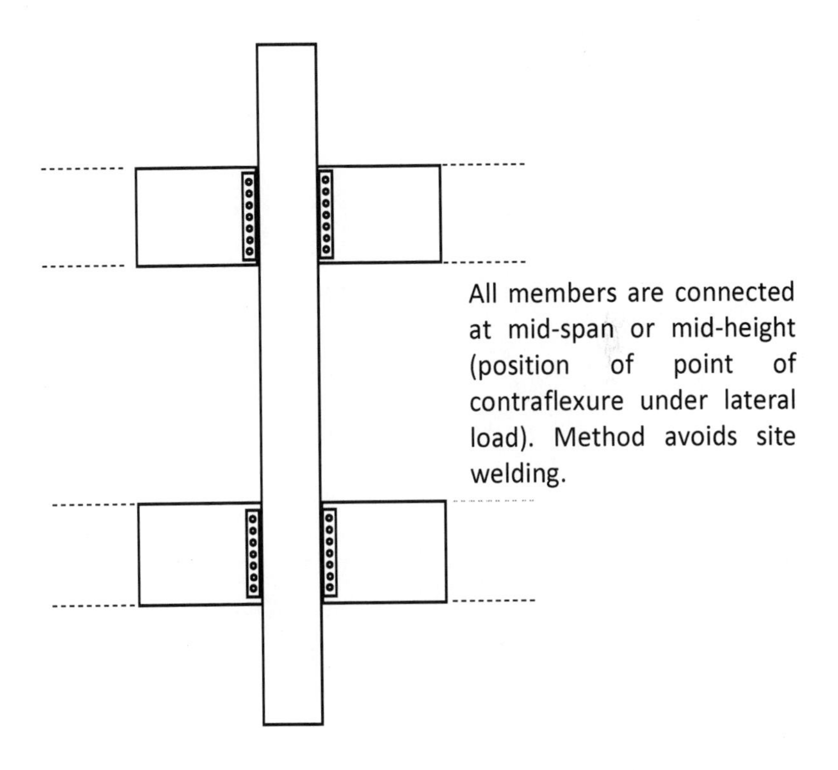

Figure 2.6: Tree" fabricated from two-storey length of column section

The latest evolution of the braced tube system is the 'diagrid' structure. Here the shear-lag problem is eliminated by triangulation of all the elements of the tube.

2.1.5 Tube-in-Tube Structures

Refers to the use of the perimeter as one tube and the core framing as the inner tube. The core is usually diagonally braced. The floors act as if pinned to the core and perimeter columns.

2.1.6 Core Interactive Structures

A special case of tube-in-tube structure. Here the spans in the outer tube are increased, reducing its shear stiffness. (Figure 2.7).

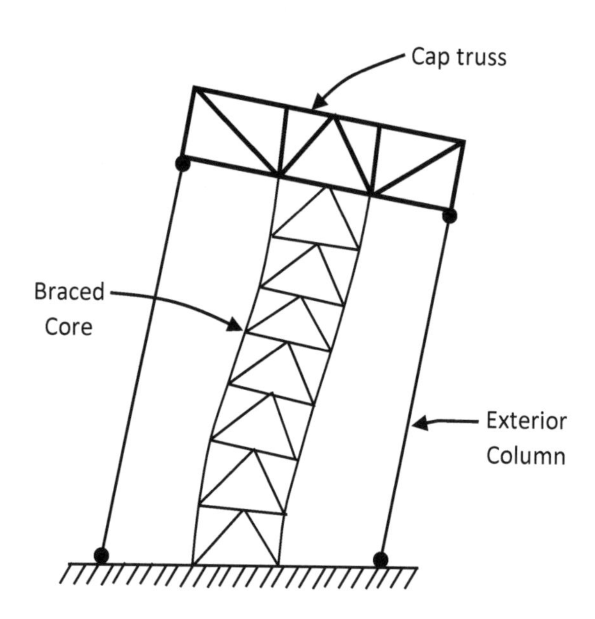

Figure 2.7: Cap truss ensures point of contraflexure at mid-height

The inner core **interacts** with the outer tube. Outrigger beams or trusses (concrete or steel) are placed at mechanical or fire safety refuge floors. These outriggers are placed from core to perimeter, unlike the belt trusses in system (d). They are popular with architects as the span between perimeter columns can be large.

The latest evolution is called "core-mast". Typically 8 façade "mega" columns resist wind with the core. Otherwise the façade has little structural role. Figure 2.8 shows the system. It is the most common modern method of lateral load resistance in very tall buildings.

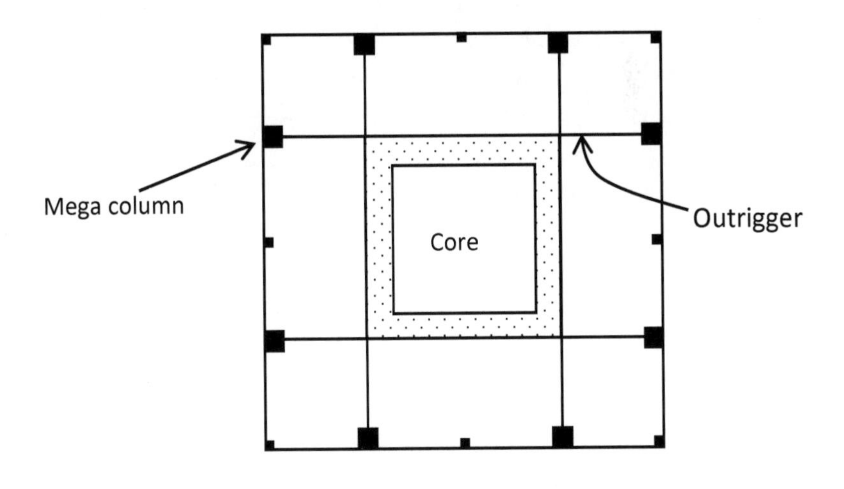

Figure 2.8: "Core-mast" system

2.1.7 Bundled-Tube Structures

Several tubes (usually nine) of various heights are connected together. Thus it reduces the shear-lag problem. Another advantage is that the system can meet the architectural demand of various floor plate sizes. It is a very efficient structure suitable for very tall structures. The weakness is the differential shortening of columns at the junctions of tubes. E.g. Willis Tower, Chicago (formerly Sears Tower).

2.1.8 Space structures ("Mega"- structures)

The system consists of a three-dimensional triangular frame structure. Being a truss, it is very efficient as lateral forces are resisted by axial forces in members rather than bending. This solves the shear-lag problem of framed tubes. Unfortunately, connections expensive if not carefully designed and detailed. E.g. Bank of China, Hong Kong.

2.2 MAXIMUM ECONOMIC HEIGHT?

The maximum economic height usually depends on the following:

1. Cost of land vs. rent;
2. Speed of construction;
3. Rentable space:

An office building up to 80-storeys typically has a core of about 25% of the plan area. Office buildings above this height may have larger cores, e.g., World Trade Centre (110- storeys) had 30%. This loss in rentable area may be unacceptable.

4. Other limits:

Traffic, e.g., Sears Tower (110-storeys) occupants (more than 10,000) queue up to two hours to leave each day.

2.3 DESIGN OF LATERAL LOAD RESISTING SYSTEMS

The main checks are:

1. Horizontal stability under lateral load:
2. Strength;
3. Stiffness.

2.4 EUROCODE 2: STABILITY CHECK ON BUILDING

Eurocode 2 requires the stability of the structure to be ensured by considering the following load case: Dead load (DL) + Wind load (WL) as shown in Figure 2.9:

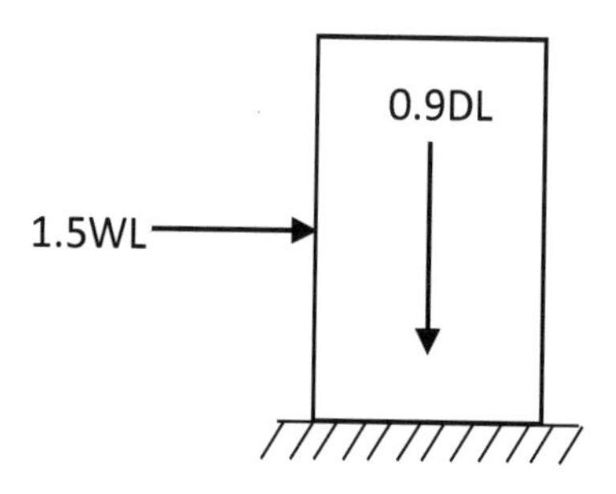

Figure 2.9: Eurocode 2 stability loadcase

2.4.1 Stability under lateral loads

Consider the building (height H) as a rigid block unconnected to a rigid foundation. Ensure that the overturning moment from the wind (w) is less than the restoring moment using the Eurocode 2 load factors. See Figure 2.10.

Thus $1.5wBH^2/2 < 0.9\rho BD^2H/2$

where ρ is the dead load density of the building (DL only).

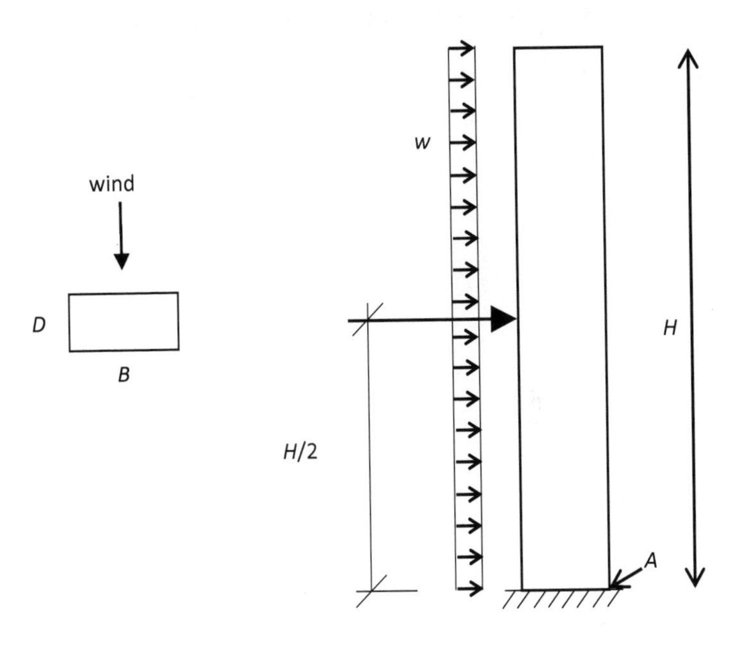

Figure 2.10: Stability of building under lateral loads

Buildings which fail this check must be held down for stability. The characteristics of buildings that fail this check: narrow in at least one dimension (*D*); slender (*H/D* high); light (low ρ); subject to strong winds (high *w*). Notice that there is no requirement for the building to be tall.

Worked example

Check the overall stability of building shown below to Eurocode 2. The characteristic (ultimate) wind load is 1.4 kN/m^2, and the building density (DL only) is 150 kg/m^3.

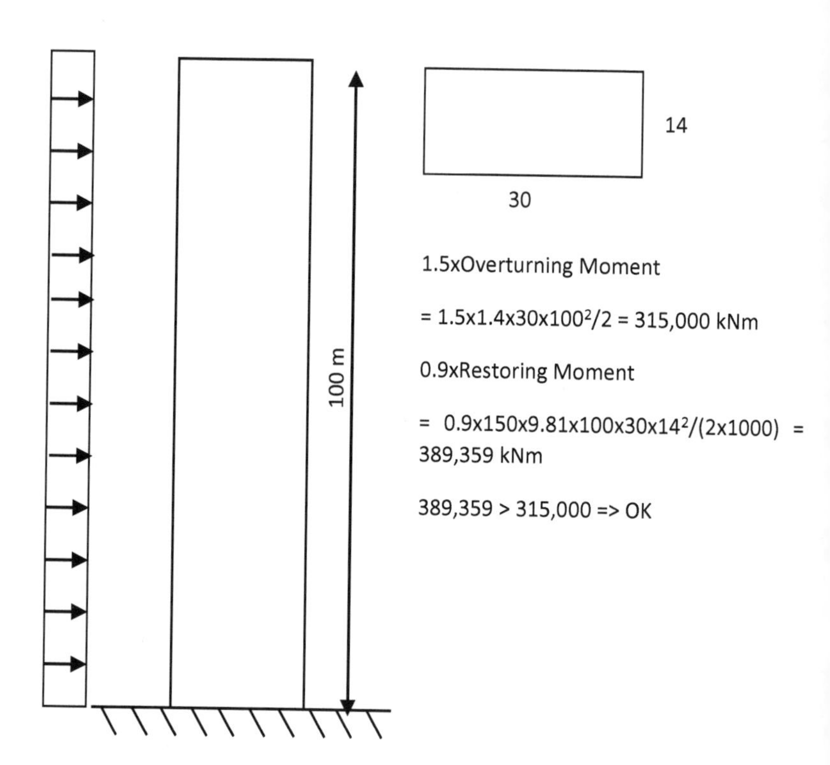

2.5 STRENGTH AND STIFFNESS

We must ensure that the elements that resist loads are strong enough. This is a straightforward matter of applying the code. For example, Eurocode 2 requires elements to be designed to resist at least 1.35DL + 1.5LL. However this is not enough. We also must limit the stiffness of the bracing system. This is to control second order effects (a safety issue) and limit accelerations (a comfort issue).

There are two second order effects (both known as P-delta effects) that we must examine:

- Due to gravity load;
- Due to lateral load.

2.5.1 Gravity Load

The critical gravity load for instability of the building is given as $P_{crit} = 7.837EI/H^2$ assuming a constant value of EI over the height and purely flexural behaviour (Timoshenko and Gere, 1961). See Figure 2.11.

We must keep well away from this load. Normal practice is to ensure the ultimate gravity load is less than $0.1P_{crit}$ An unacceptable loss in stiffness, and so capacity, takes place if the axial load exceeds this load. The magnification factor for moments is $1/(1-P/P_{crit})$. So for $P/P_{crit} = 0.1$ this means moments are magnified 11%.

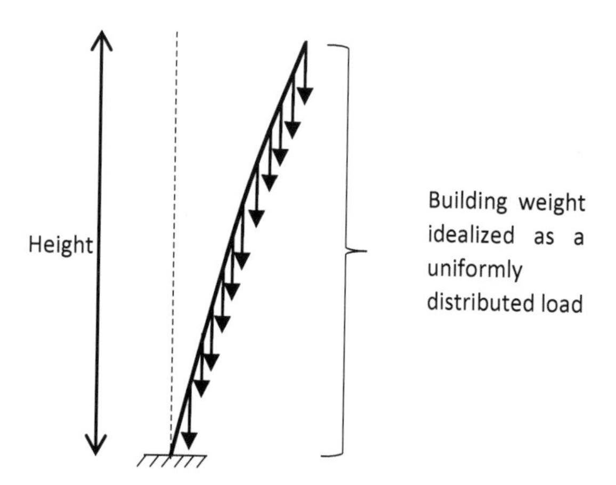

Figure 2.11: Overall stability failure of building

2.5.2 Lateral Load

Regardless of strength, the building deflects under lateral load. Once it deflects the line of action of the gravity loads moves. This increases moments on the foundation. Thus the frame must be stiff enough so it deflects less and these second order effects (P-delta effects) are small.

Consider a building subject to a uniformly distributed lateral load whose resultant is *P*. A first-order analysis (i.e. where the deflections are assumed to be small enough to ignore) of the building gives an estimate of the base moment, say $M_1 = PH/2$. If it is assumed that the deflected shape is linear (a reasonable guess for a tall building), then including the resulting movement of the building weight, the total moment at the base is larger by $W\Delta/2$. Thus the Modification Factor (*MF*) is always greater than 1.0. See Figure 2.12.

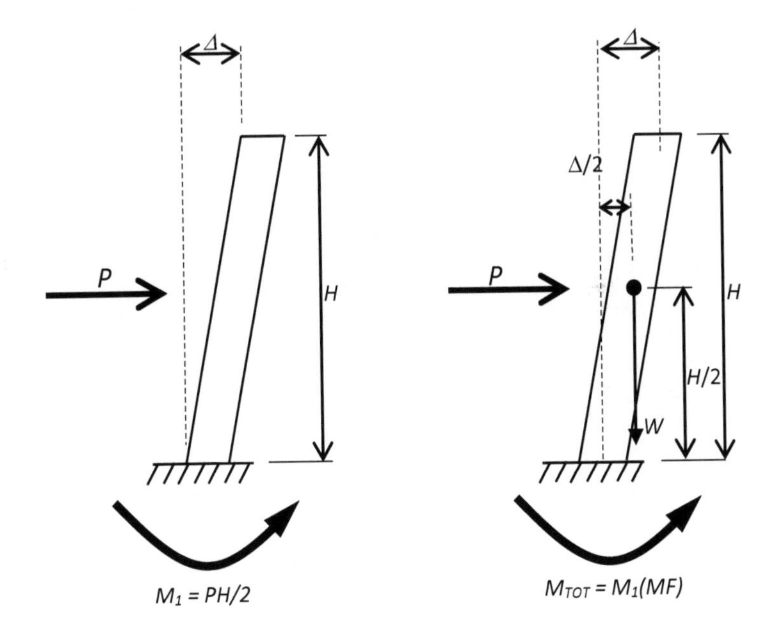

Figure 2.12: Base moment from first order (left) and second order (right) analysis

Thus the first order analysis ignores the effect of the deflection on the base moment. Second order analysis includes this effect.

Worked Example

Following 3-storey car park is to be used in Singapore. Floors are 250 mm thick flat plates.

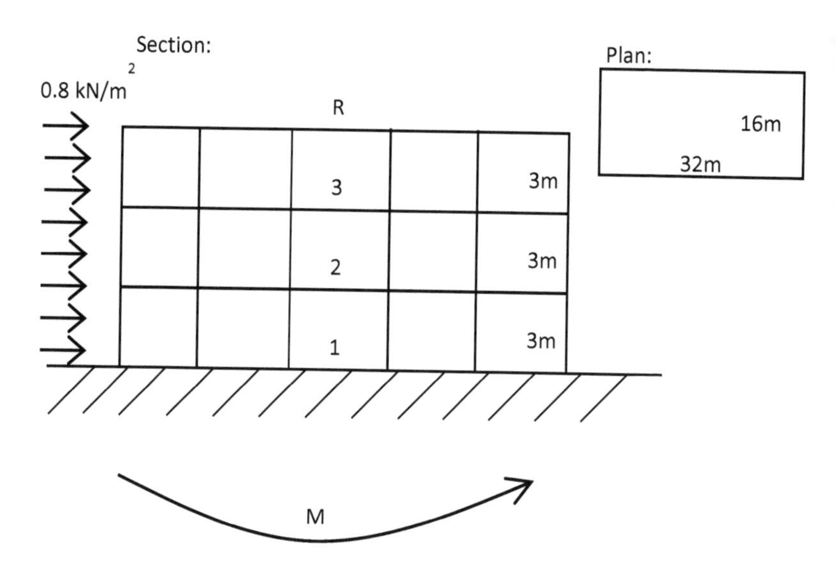

Neglecting any other weight except floors (e.g. cladding, columns), estimate the **modification factor** (MF) in the long direction of the building due to lateral load second order effects. Assume the frame is designed to move

H/500 at roof under a wind load of 0.8 kN/m^2. In addition, assume the deflected shape is *linear* and concrete density is 25 kN/m^3.

Solution

Ignoring deflection: *M* at foundation = 0.8x16x9x4.5 = 518 kNm

Linear deflected shape implies that if the roof moves *H*/500 = 9000/500 = 18 mm, level 3 moves 12 mm, and level 2 moves 6 mm.

Thus additional moment:

= (0.25x25)x32x16x(0.018+0.012+0.006) = 116 kNm

Hence total moment foundation must resist is 518 + 116 = 634 kNm. Thus MF = 634/518 = **1.22**

This MF is excessive. The stiffness of the lateral load resisting system should be increased so that the MF is not greater than about 1.1. It can be shown that this means deflection should be not greater than about 11mm at the roof.

2.5.1 Approximate formula for MF

The following approximate formula derived by Robertson (1987) evaluates modification factor (moment magnifier) for a given drift ratio (i.e., deflection at top of building). It is suitable for preliminary design of multi-storey buildings higher than 10-storeys or so.

Modification Factor Formula

Modification Factor (MF) = 1/(1 – WR/Q)

where

"*W*" is building weight (in Newtons),

"*R*" is drift ratio (e.g., 1/500), and

"*Q*" is total wind load (in Newtons).

The above formula is derived using five main assumptions: Structure is prismatic; Deflected shape is linear; Mass of building is distributed uniformly (i.e., centre of mass of building at mid-height); Building resists a uniform wind pressure; Foundation is rigid.

If the modification factor is excessive (above 10%) then the drift should not be allowed to reach this ratio, i.e., frame must be made **stiffer**.

2.5.2 Tall building with rectangular plan:

If the plan is rectangular, the formula demonstrates MF is biggest in the *long* direction. This is because in the long direction the wind forces will be relatively small as the width of building is small. Thus, in the long direction, the foundation will not be designed for a large moment. So, as a percentage, the addition in foundation moment due to P-delta effect is large, as foundation was designed for a small moment in first place.

Illustration: consider a building of aspect ratio of 2 on plan. Suppose the wind causes a base 1st order moment (M_1) of 100 units when the wind is blowing on the broad side of the building, and this results in an additional moment (M_{ADD}) of 10 units. Now when the wind is blowing onto the narrow side of the building M_1 is 50 units. Assuming the frame in both directions is designed for equal drifts, then M_{ADD} is also 10 units.

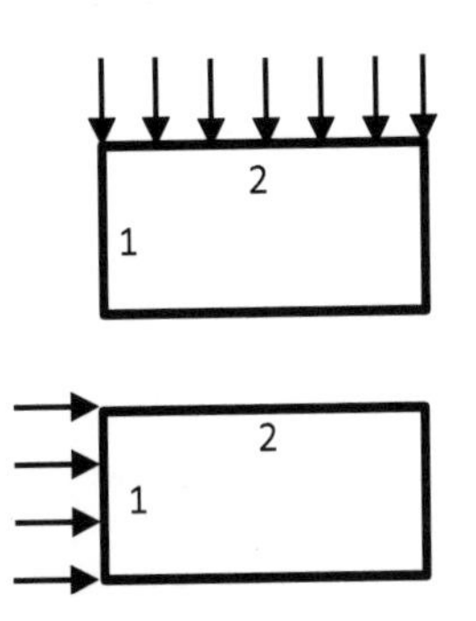

Moment at base

M_1	M_{ADD}	M_{TOT}	MF
100	10	110	1.1
50	10	60	1.2

Hence the modification factor (M_{TOT}/M_1) is larger in the long direction of the building.

Worked Example using formula:

Hypothetical 30-storey concrete building in low wind environment, e.g. Singapore/Malaysia, designed using drift ratio *H*/500. Calculate *MF* in long direction.

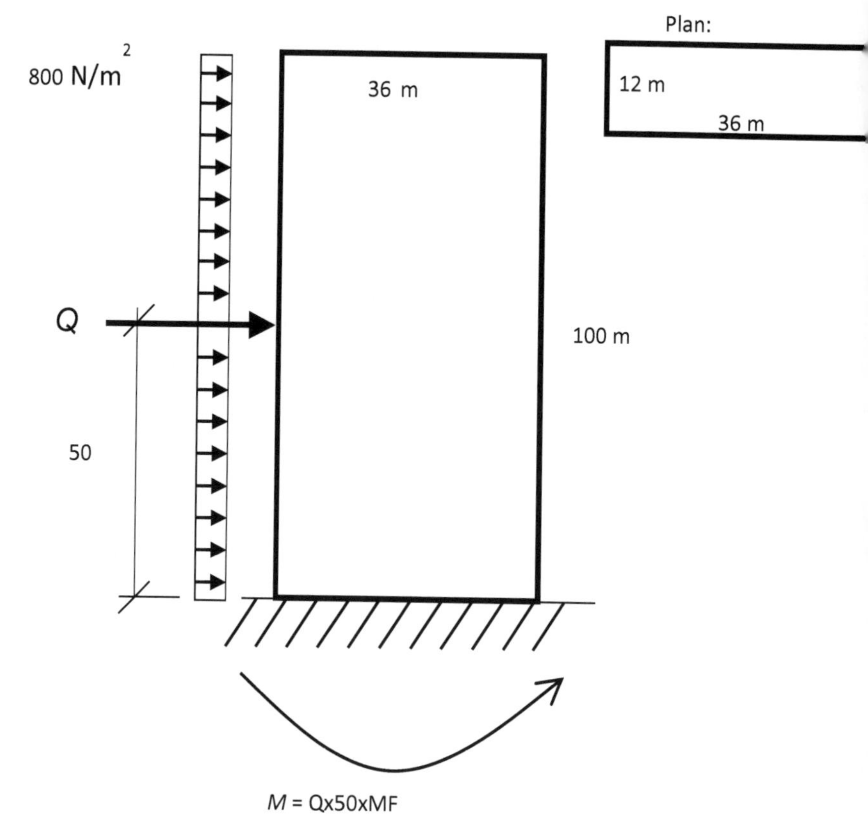

Solution:

Wind force on narrow face: Q = 800x12x100 = 9.6×10^5 N

Building Density = 400 kg/m^3 (typical for a concrete building)

W = 400x12x36x100x9.81 = 1.7×10^8 N

MF = 1 / (1-WR/Q)

= 1/ (1- 1.7×10^8/500x9.6×10^5) = 1.55

For a given wind load intensity MF does *not* depend on building height. We should always check this P-Delta effect for every building, especially a tall building (as the consequences of failure are greater).

Notice to minimize MF we must minimize WR/Q (e.g. WR/Q = 0.5 implies MF = 2 and WR/Q = 0.1 => MF = 1.1). Thus the commonly used drift ratio of 1/500 limit may **not** be appropriate for a low wind load environment like Singapore or Malaysia. (Small value of wind Q means larger MF. Small Q for the same R implies a more flexible frame.) (In fact, CIRIA report 107 recommends that lateral deflection of multi-storey buildings should not exceed Height/1000 (based on one in 50-year wind). Examples in chapter 6 show that when the wind is low the required lateral deflection can be even lower, i.e. the allowable drift is less).

In addition, notice too that this formula is telling us to be careful if floor plates are large and especially if their aspect ratio (on plan) is large as well.

Be especially careful of concrete high-rises whose density is 400 kg/m^3 i.e., about double that of steel building (Schueller, 1986). Thus they must not be allowed to deflect excessively.

2.5.3 Occupant comfort

This is often a critical factor affecting the lateral load system stiffness. The current accepted (but not codified) acceleration limits are based on a wind with a return period of 10-yrs: (Taranath, 1998).

- Office: max acceleration = 0.25 m/s^2.
- Apartment/Hotel: max acceleration= 0.15 m/s^2.

Note: Wind load with a return period of 10-years (used for acceleration calculations) is about **20% less** than that for 50-years.

The following table shows likely accelerations in various situations (Table 2.1). Clearly any limit on acceleration must take into account the likely duration. Such limits are subjective and vary greatly between individuals. These are the main reasons such limits are rarely codified.

Table 2.1: Accelerations in various situations

Situation	Acceleration (m/s^2)	Duration (s)	Reaction
Diving Jet	55-70	1	Special training required
Rollercoaster	45	8 peaks in 270 s	Exciting, popular
Buildings during major earthquakes	20-45	0.1 repeated	Horrifying
Buildings during storms	0.05-1	1,000+	Nauseating, alarming, unacceptable

2.5.3.1 *Artificial damping*

Adding stiffness to the frame actually has little effect on accelerations. Should we add dampers? It is true that dampers reduce accelerations and also deflections. However, they are often expensive and troublesome to retrofit: e.g., John Hancock Tower, Boston (see case study). They may also be a maintenance item although modern systems are virtually maintenance-free, e.g., sloshing-water type or pendulum type.

At the preliminary design stage accelerations are difficult to predict. Space may be left near or at roof level in borderline cases (Grossman 1990). Accelerations are rarely a problem with concrete buildings. The strategy usually used is to expect damper to be required only on very tall (more than 250 m), slender buildings (SR more than 5).

Although an artificial damper will reduce the deflections and so the lateral load P-delta, it should not be considered a safety device: the safety of the building should not depend on the supply of electricity! Thus use it to solve a **comfort** problem rather than a **safety** problem (i.e., P-delta). Thus have enough stiffness from the bare frame to control P-delta effects. Only in the case of very tall, slender and light buildings (often steel) this is usually enough to control accelerations too.

Many of tallest and most slender high-rise buildings in world do not need damping for accelerations to be acceptable, e.g. Burj, Dubai, WFC Shanghai.

2.6 SOURCES OF ERROR IN CALCULATIONS

2.6.1 Rotation of base of building

Routinely assumed as fixed but elastic deformation of piles/caissons and creep of soil will mean more rotation, increasing lateral load P-delta effects. Model using spring elements in detailed design stage.

2.6.2 Cracking/creep of RC elements (resulting in an increase in drift)

Reduce I of horizontal members (beams and slabs) by 50%. Reduce E of concrete by 20%. (Stafford-Smith and Coull, 1991).

Thus preliminary design should not be near the limit for these P-delta requirements.

CHAPTER 3: CONCRETE BUILDINGS

3.1 SHEAR WALLS

3.1.1 Eurocode 2 formula

Concrete buildings are often "braced" by shear walls (i.e., lateral load is resisted by walls). In addition, we usually also require them to be "non-sway" (i.e., second order effects are small.

According to Eurocode 2 shear walls will be sufficiently stiff if proportioned according to this expression:

$$\sum E_{cm} I_c \geq F_v(n + 1.6)\, H^2/0.517n$$

where

F_V = total vertical load on the whole structure at ultimate;

n = number of storeys;

H = height of building;

E_{cm} = mean value of modulus of elasticity of concrete;

I_c = second moment of area of **uncracked** concrete wall (if the wall is likely to be cracked then use 0.517/2 instead of 0.517 above, i.e. requirement is doubled).

The derivation of this formula appears in Annex H of Eurocode 2 and assumes that: a) the layout of structure on plan is reasonably symmetrical; b) the wall stiffness is reasonably constant throughout height c) shear walls possess no large openings; and d) the base is fully fixed. The formula is based on overall stability considerations under gravity load only. Such considerations of buckling resistance require a certain value of second moment of area to resist the buckling and are often a critical consideration given that concrete buildings are typically twice as heavy as their steel counterparts (Schueller, 1986). Use of this latter formula usually ensures second order effects will be low. Thus it ensures the frame is braced and non-sway.

As seen in the previous chapter, $P_{crit} = 7.837EI/H^2$.

From Eurocode 2: $F_V < 0.1P_{crit}$

Also $EI \approx 0.66\ E_{cm}I_c$ for *uncracked* sections

And $F_V = 0.1\xi \sum EI/H^2$

where $\xi = 7.8(n)/(n + 1.6)$ assuming bracing members have constant stiffness over the height, the vertical load is approximately uniformly distributed, and that there is a rigid base.

Thus $\sum E_{cm} I_c \geq F_v(n + 1.6) H^2/0.517n$

Eurocode 2 recommends how this expression should be modified when the base is not fixed and also when shear deformations are not small.

We now have two independent requirements for bracing stiffness:

1) Eurocode 2 formula for gravity stability second order effects;

2) MF formula for control of lateral load second order effects, noted in the previous chapter.

The requirement of 1) leads to the provision of equal stiffness in both directions regardless of the plan shape, while 2) requires more stiffness in the long-direction of the plan. As can be seen from the examples in chapter 6, the requirements of 1) often mean 2) is automatically satisfied.

Worked Example using Eurocode 2 formula:

The following concrete building is 20 m x 20 m on plan. If it is 30-storeys high, what shear walls are necessary according to Eurocode 2? Assume concrete f_{cu} = 45 N/mm^2.

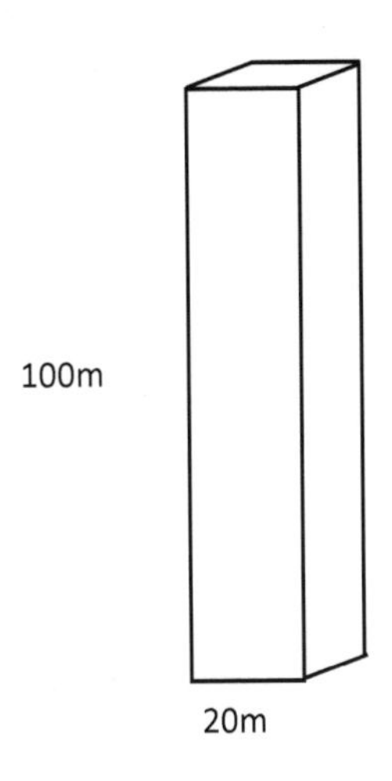

Solution:

Weight of typical concrete building = 400 kg/m^3

Assume this figure includes lightweight cladding.

Thus, using γ = 1.5,

F_v = 400x20x20x100x1.5 = 24x10^6 kg = 24x10^{6}x9.81 N

= 235x10^3 kN

f_{cu} = 45 N/mm^2 and E_{cm} = 34x10^6 kN/m^2

Eurocode 2 formula: $\sum E_{cm}I_c \geq F_v(n + 1.6)\, H^2/0.517n$

$\Sigma E_{cm} I_c$ > $235x10^3$ $(30+1.6)100^2/(0.517x30)$ > $4788x10^6$ kNm^2 ,Thus $\Sigma I_c > 4788x10^6/34x10^6 = 141\ m^4$

Try a central core, say 9 m square with 400 mm walls (less than 25% area):

$I_{xx} = BH^3/12-bh^3/12 = 9x9^3/12-8.2x8.2^3/12$

$= 170\ m^4 > 141\ m^4$ => okay

- Thus using central core:

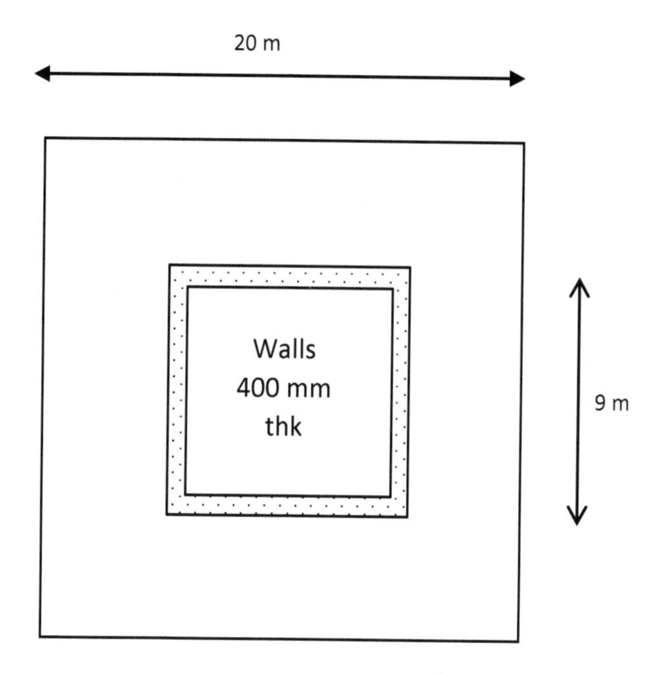

3.1.2 Cracking

It is appropriate to use the full value of second moment of area I to estimate deflections as long as the particular wall remains uncracked. Thus the walls should be loaded with as much gravity load as possible to help ensure they remain uncracked.

3.1.3 Layout of shear walls on plan

The following (Figure 3.1) shows characteristics of various layouts.

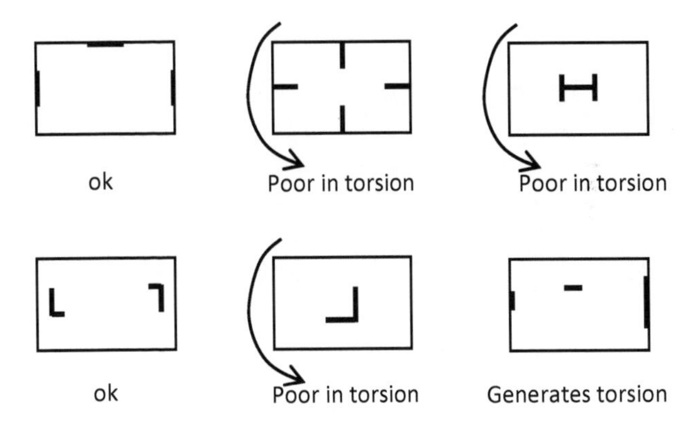

Figure 3.1: Layouts of shear walls

We should have at least three shear walls, two of which should be symmetrically placed and equal in size. (The Eurocode 2 formula apparently requires we have four walls symmetrically arranged so that there is little torsion). Otherwise **torsions** may be large. Ideally, we want the centre of resistance (shear centre) and the centre of the applied load to coincide.

3.1.4 Unsymmetrical layout of walls

Consider the following arrangement of walls:

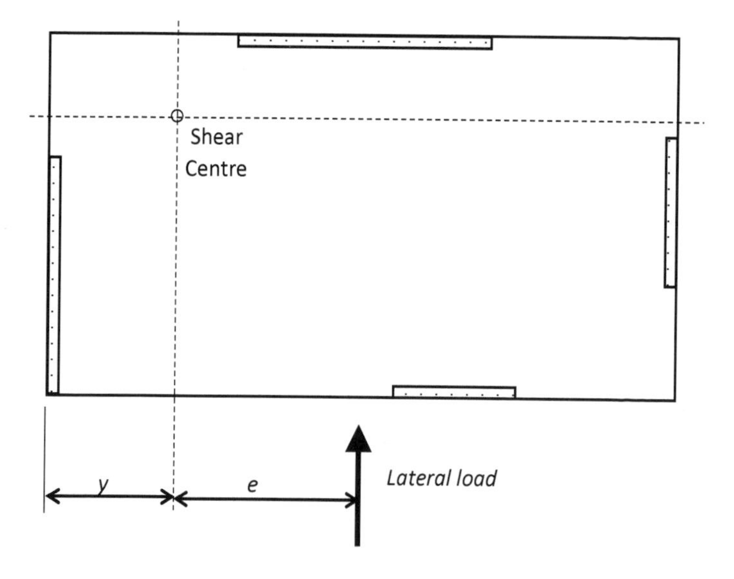

By analogy with the centroid, the position of the shear centre (centre of stiffness):

$$y = \sum I.y / \sum I$$

Where I is the second moment of area of the wall about an axis perpendicular to the lateral load.

Worked Example (Broker, 2006)

Find the torsion generated by the layout of walls below when a wind of 1 kN/m² blows. Wall $w_3 = w_4$. Assume the thickness of all walls is 0.2 m and the storey height is 3.6 m.

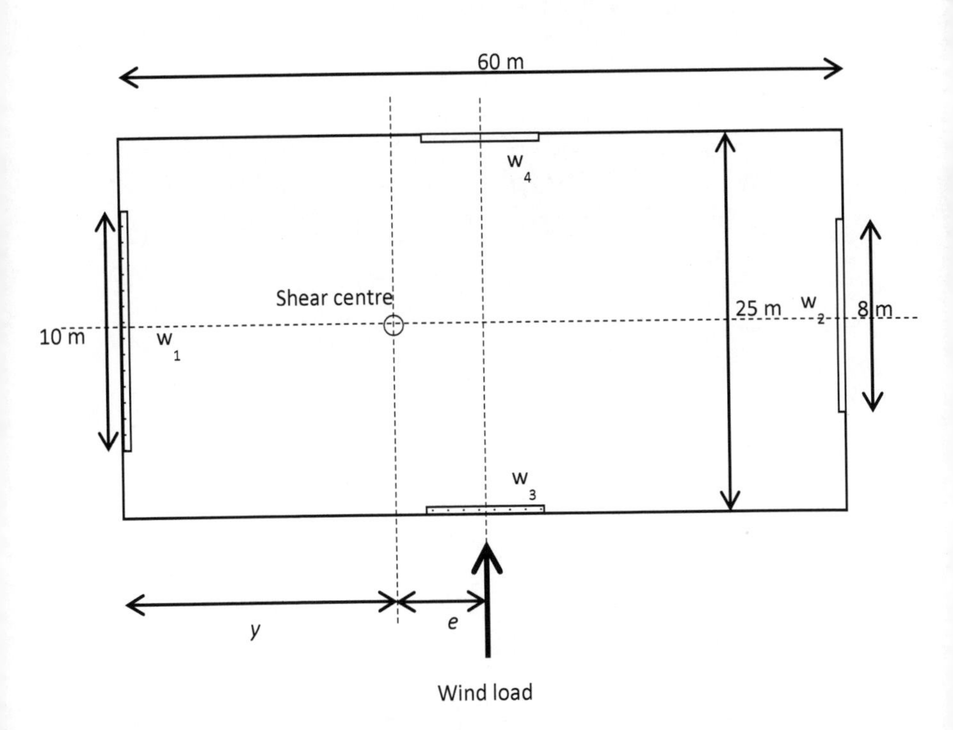

The centroid of stiffness $y = \sum I\, y/\sum I$

Wind blowing is resisted mostly by w_1 and w_2 (compare values of second moment of area to see resistance of w_3 and w_4 negligible).

The relative stiffness of the walls are as follows (all equal width so width and 1/12 terms cancel):

w_1: 10^3 = 1,000 m^3 and w_2: 8^3 = 512 m^3

y = (1,000x0.1 + 512x59.9)/(1,000 + 512) = 20.3 m

Eccentricity, e = 30 – 20.3 = 9.7 m.

Twisting Moment, M_t = 9.7x1.0x60x3.6 = 2,095 kNm per floor.

This is a large torsion (if resisted by w_3 and w_4 only, the force induced in each is 2,095/24.8 = 84.5 kN. Compare this with the force they would take if the wind blew perpendicular to the wind direction shown, i.e. 25x3.6x1.0/2 = 45 kN).

Thus an unsymmetrical layout of walls can generate torsion.

We will consider the distribution of lateral load between walls in more detail in the next section.

3.1.5 Distribution of load:

Consider, for example, the layout below. Notice that it is symmetrical.

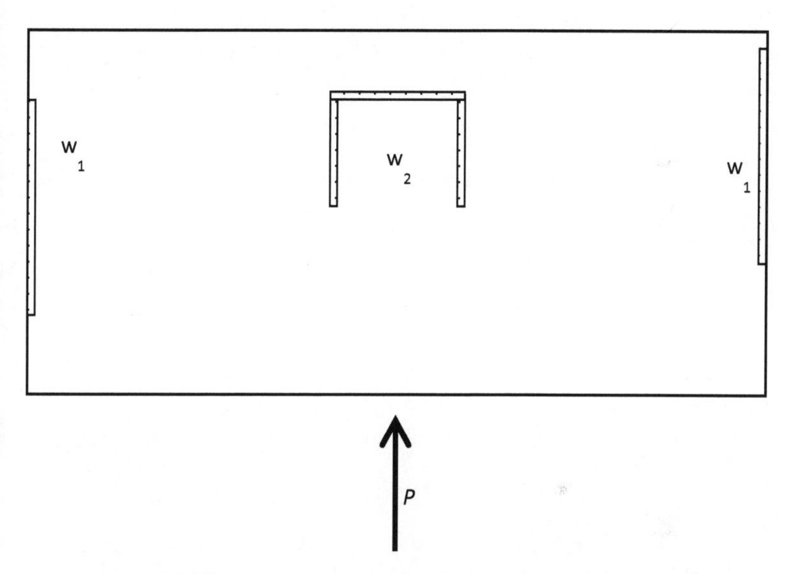

To model the behavior a rigid beam on spring supports should be considered (rather than the elastic beam on rigid supports).

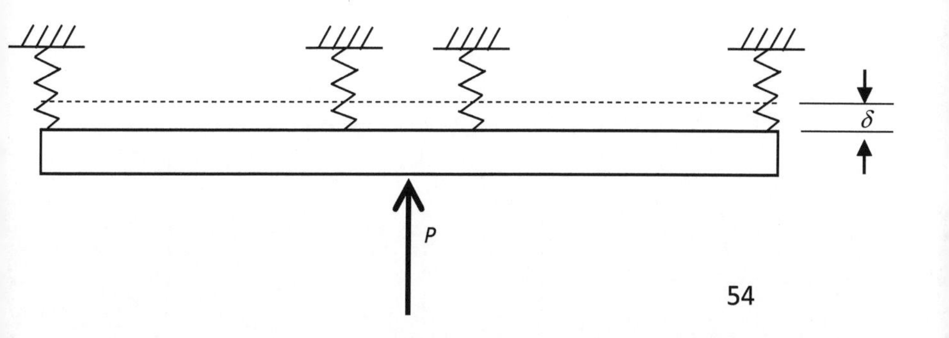

In most cases the assumption that the floor diaphragm is rigid is a reasonable one. For a concrete slab this is likely to be true unless there are large openings in the slab. Thus the walls will deflect the same amount, δ.

Consider two walls each loaded by their share of the total lateral load P, namely P_1 and P_2 where $P = P_1 + P_2$. Also $I = I_1 + I_2$.

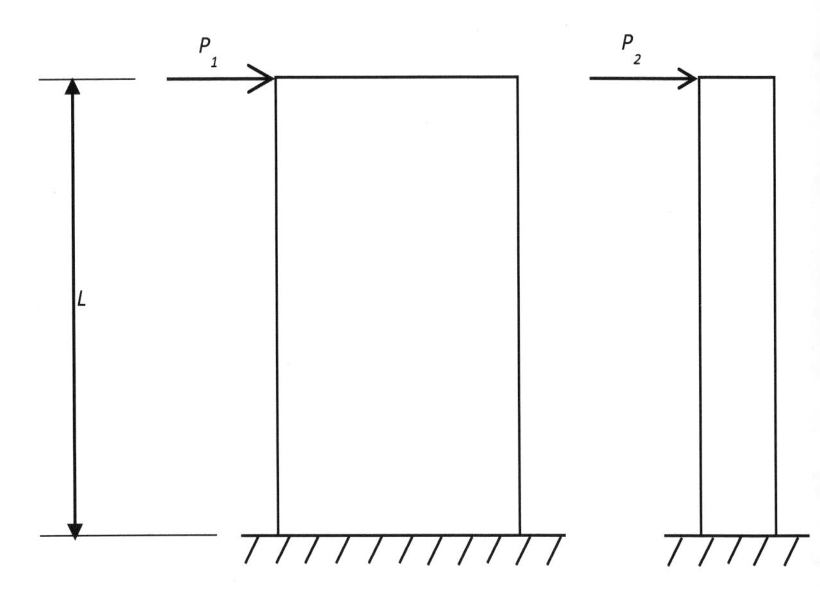

$\delta = P_1L^3/3EI_1 = P_2L^3/3EI_2$

and $\delta = PL^3/3E\Sigma I$, thus $P_1 \alpha P I_1/\Sigma I$ and $P_2 \alpha P I_2/\Sigma I$

i.e., Distribution of load to shear walls is according to the second moment of area. (As we would expect in a statically indeterminate structure, the stiffer element attracts more load.)

The usual assumption at preliminary design stage is that walls take entire lateral load. However, if the height of the structure (and of course the shear walls) is greater than about 25 m, then **interaction** occurs between walls and frame. It thus becomes less accurate to assume all of lateral load is resisted by walls.

Worked Example:

Calculate distribution of wind load in y-direction. Walls are 200 mm thick. Wind load is 1 kN/m^2. Storey height is 3 m.

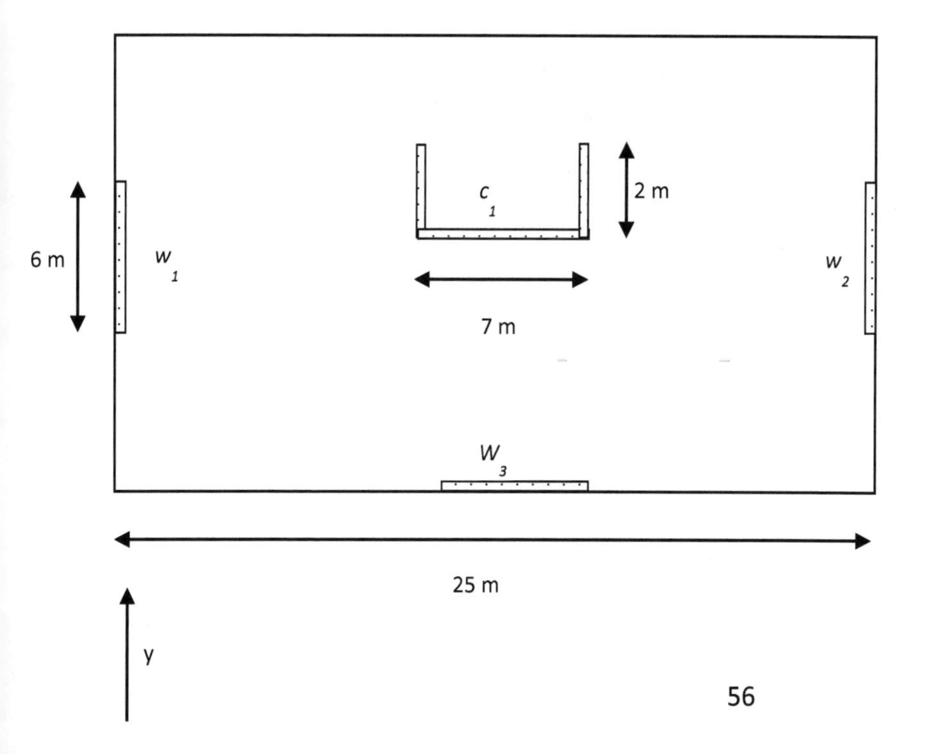

Approximate value of second moment of area (Bourne 2013). Assuming a thin-walled section of arbitrary shape:

- Second moment of area, $I = 0.55Ay_{bot}y_{top}$

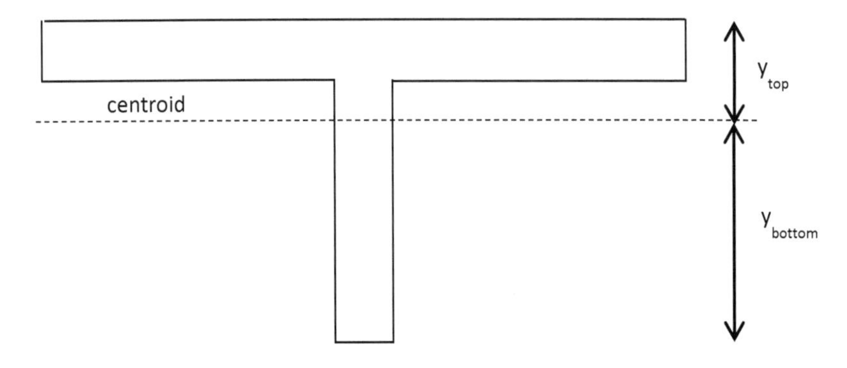

Applying this to our case:

Section properties: consider "C_1"

Area, A = 0.2x1.8x2+7x0.2 = 2.12 m^2

Distance to centroid, y_{bot} = (2x0.2x1.8x1.1+7x0.2x0.1)/ 2.12 = 0.44 m

Second moment of area, I_y = 0.55x2.12x0.44x1.56

= 0.8 m^4

Wall thickness, 0.2 m

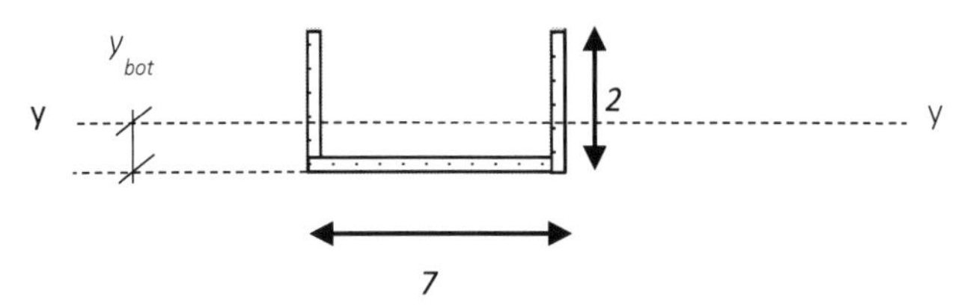

Properties of w_1 (w_2 similar): $I_y = (1/12)(th^3) = (1/12)(0.2x6^3) = 3.6\ m^4$

Properties of w_3 : $I_y = (1/12)(ht^3) = (1/12)(6x0.2^3) = 0.004\ m^4$ (small, so neglect)

If uniformly distributed wind load of 1 kN/m^2 blows in y-direction:

Wind load per each floor = P = 3x25x1 = 75 kN.

Load taken by $w_1 = P.I_{w1}/\Sigma I$ = 75x3.6/(3.6+3.6+0.8) = 34 kN, i.e., 45% of total wind load.

Similarly load taken by w_2 is 34 kN and therefore load taken by core c_1 is 7 kN.

These loads are transferred from the cladding to the walls through the floor which acts as a diaphragm.

Wall w_1 designed for 34 kN at typical level.

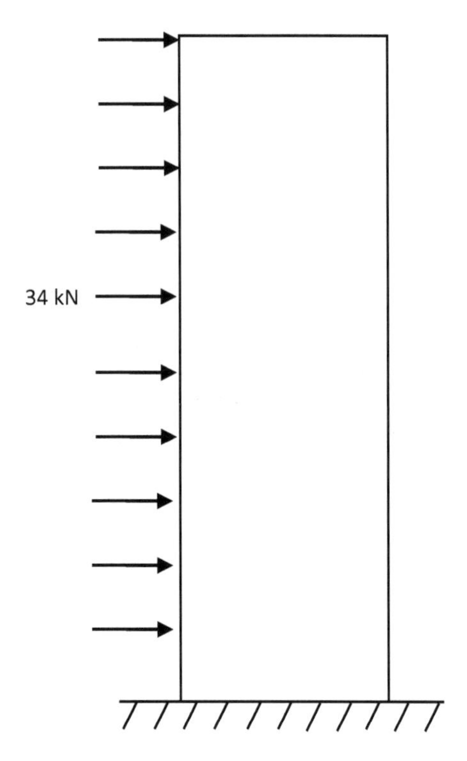

We can use elastic theory to analyze the wall under the effects of the lateral and axial load (we must ensure that the wall is pre-compressed by gravity loading).

Worked Example:

Calculate the stresses at ultimate in this 200 mm thick wall. Ultimate loads are as shown.

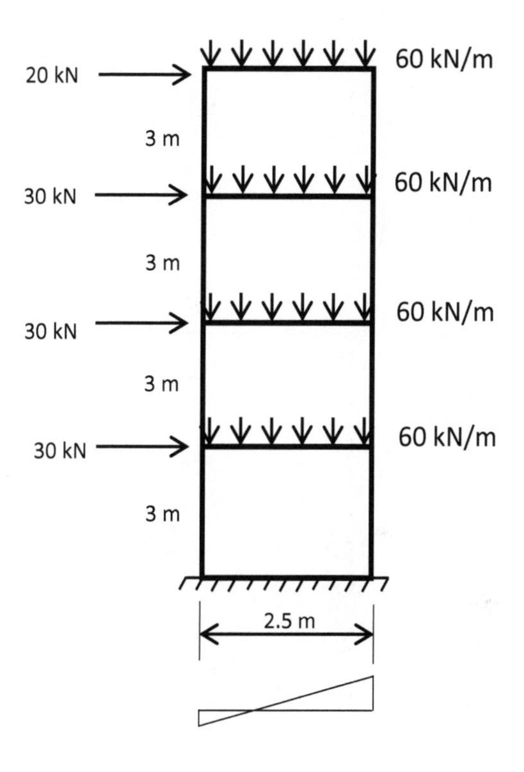

N = 60x4x2.5 = 600 kN; M = 20x12+30x(9+6+3) = 780 kNm

$f_t = N/(Lt) +/- M/(tL^2/6)$

$= 600{,}000/(2500\text{x}200) +/- 780\text{x}10^6/(200\text{x}2500^2/6)$

= 1.2 + 3.7 = 4.9 N/mm^2 compression and 1.2 – 3.7 = - 2.5 N/mm^2 (i.e. tension)

Assume that the tension is resisted by 1 m at the end of the wall.

$A_s = 0.5f_tL_tt/(0.87xf_{yk})$

$= 0.5x2.5x1000x200/(0.87x500) = 575\ mm^2$

Design remainder of wall as columns supporting compression max 4.9 N/mm^2.

3.1.6 Holes in shear walls: "coupled walls"

Openings should be placed near the central part of the wall (neutral axis) and staggered if possible. The values of relative stiffness that can be assumed are shown in Figure 3.2. (CIRIA 1984)

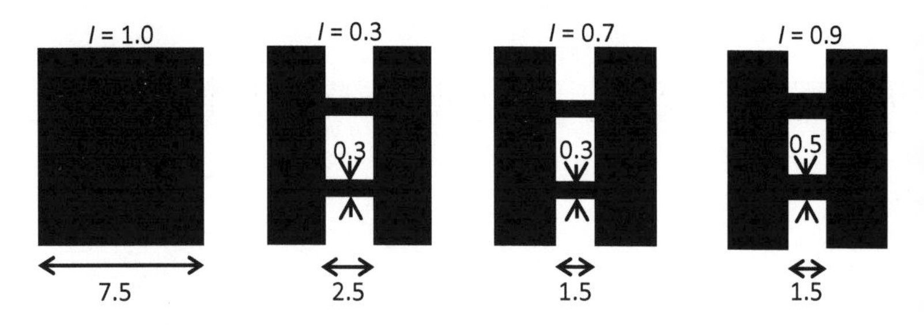

Figure 3.2: Stiffness of walls with openings

3.2 Concrete Buildings: Gravity systems

Lightness of the system chosen is important for the foundation, if transferring loads and to lessen the requirement for walls.

Systems using *insitu* reinforced or prestressed concrete:

- Flat plate.
- Flat slab with drops or column heads.
- One-way beam and slab.

Systems using hybrid construction (precast plus *insitu*):

- One way beam and slabs (e.g., Holocore).

The following recommended depths for post-tensioned elements is from TR43, 2005 (Table 3.1):

Table 3.1: Post-tensioned concrete: floor thicknesses:

Floor Types	*ADL+LL* (kN/m^2)	Span/Depth (6m-13m)
Flat Plate	2.5	40
	5.0	36
	10.0	30
Flat Slab	2.5	44
with drops[1]	5.0	40
	10.0	34
One-way slab	2.5	45/25
with wide	5.0	40/22
beam[2]	10.0	35/18
One-way slab	2.5	42/18
with narrow	5.0	38/16
beam[3]	10.0	34/13

Notes to table:

1. Drop panel length ≥ span/3, depth ≥ $3h/4$ where h is the slab thickness.

2. Wide Beam: typical width of beam = span/5

3. Narrow Beam: typical width of beam = span/15

Once the decision is made to use post-tensioned slabs we should check that the layout of shear walls does not impose large restraint to creep and shrinkage of the slab. This is shown in Figure 3.3.

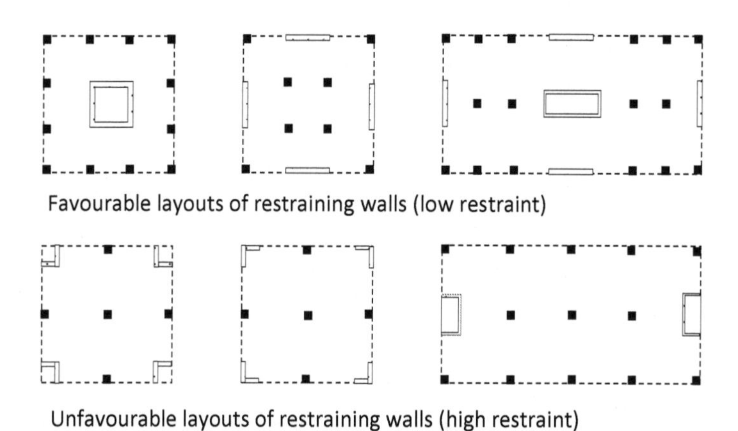

Figure 3.3: Layout of shear walls if floor post-tensioned

3.3 Axial load in columns

Use the 'tributary area' on each floor to allow the column loads to be estimated. This means a column is considered loaded by an area equal to half the span of the floor each side of the column. See Figure 3.4.

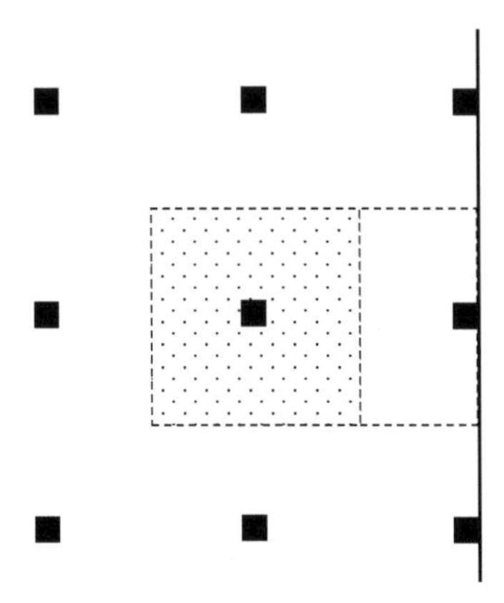

Figure 3.4: Tributary method

An additional 10% should be added to each axial load to account for uncertainties. It should be noted that the final design, which may be based on an elastic analysis, say by FEA, will produce a set of reactions different from the above. Bear in mind that these reactions may not represent a 'more accurate' set of loads, as the true column reactions are likely to be "unknown and unknowable" (Heyman 2008).

3.4 Transfer Beams

Columns at relatively close spacing may be appropriate for upper floors, but usually an open lobby area is required at the ground floor level.

These transfer structures are usually beams or plates. The structures are sometimes post-tensioned. Since they are usually loaded by columns, shear is usually critical.

In the detailed design, it is important to take account of construction procedure as the loads to transfer structures are usually larger.

Why is post-tensioned concrete particularly suitable for transfer beams?

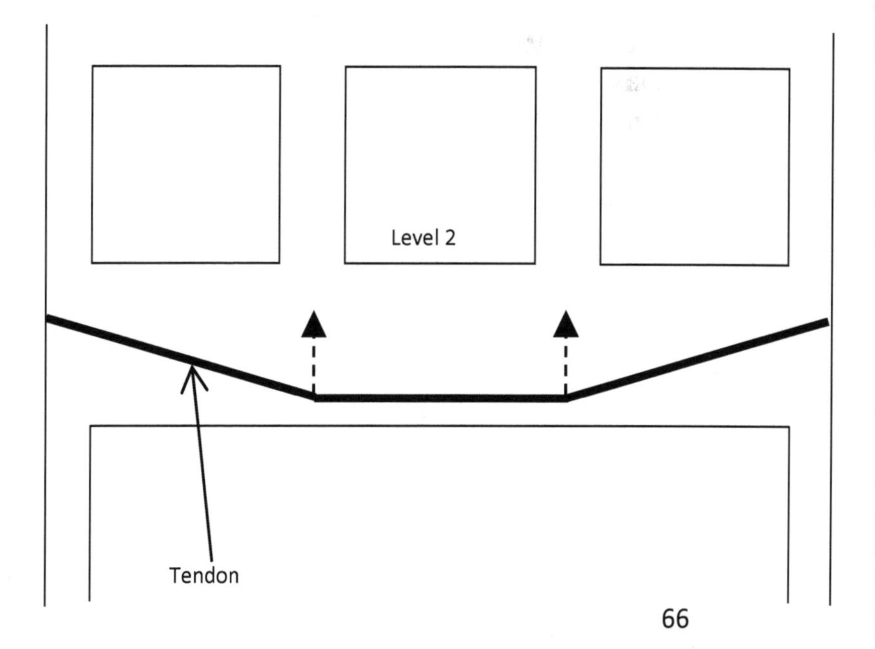

Much of the load is dead load, so this load can be 'balanced' i.e. counteracted, by upward loads from post-tensioning. Thus the beam will experience no bending from this load. As the upper structure is vulnerable to relative vertical movement of its supports, post-tensioning is advantageous because it allows deflections to be controlled/substantially reduced.

3.5 Use of High Strength Concrete

It is common to use a higher concrete strength for columns than for floors. Using normal construction methods means that the floor is placed using concrete of lower strength. ACI code (ACI: 318:14) allows a column to have a strength 1.4 times that of slab with no change in construction procedure required (it is considered that portion of column within floor is restrained, thus stronger). Thus, if floor f_{ck} = 40 N/mm^2 column can have f_{ck} = 55 N/mm^2 with no change in normal construction procedures.

When the ratio of the strengths exceeds 1.4 there are two possibilities:

(a) Place high strength concrete in the column area of the floor (Figure 3.5 (a)).

(b) Alternatively, use lower strength concrete in the floor and place extra compression reinforcement to compensate (Figure 3.5 (b)).

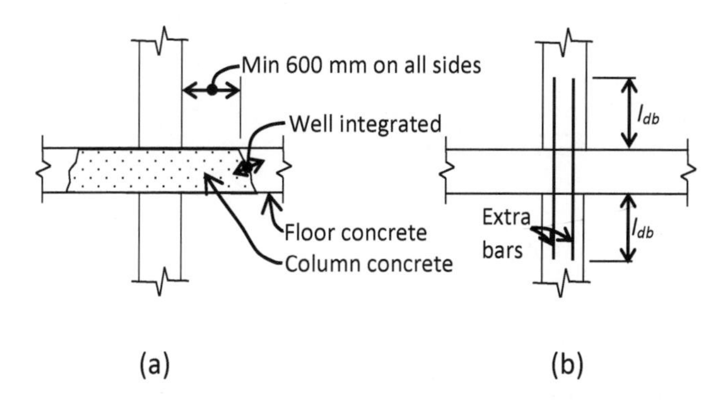

Figure 3.5: (a) High strength floor or (b) extra reinforcement.

3.5.1 Comparison of columns

The following illustrates various column sizes giving the same axial capacity (Eisele, 2005).

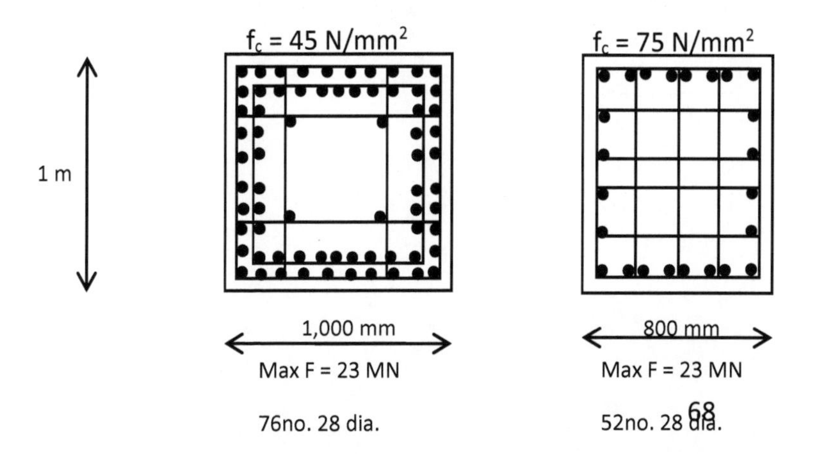

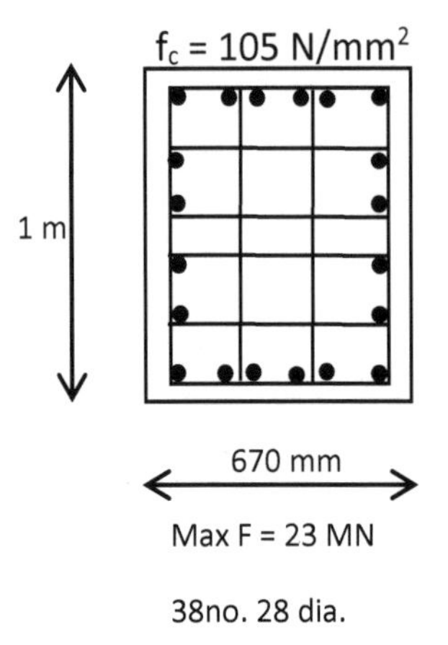

Max F = 23 MN

38no. 28 dia.

3.6 CONSTRUCTION OF SHEAR WALLS

Low-rise concrete walls constructed in the traditional fashion (i.e. floor-by-floor) are considered slow. High-rise walls are usually constructed using slip-form or jump-form methods. The choice is often based on price and/or contractor preference (there is little difference in speed). But jump-form systems are generally preferred in Singapore/Malaysia as they usually allows floors within the core to be constructed with the walls. Usually the walls are constructed at least 4 floor levels ahead of the floor (to avoid the walls being on the critical path). The floor structure is often connected to the walls using pockets (thus pinned connections only). Slip-form has the disadvantage that it requires continuous concrete supply.

3.7 COST SAVING IN CONCRETE FRAME

The cost of **formwork** is about 50-60% total cost of structural frame. (Fling 1987). With this in mind, consider ways to simplify formwork use.

e.g., space columns uniformly to allow flying forms to be used; use the same column section and wall thickness through 7-8 levels (vary reinforcement and/or concrete strength); make beam sizes standard if possible; avoid suspended formwork e.g., depressions in tops of slabs; where possible avoid pilasters, downstands, haunches, drops etc.

3.8 ADVANTAGES OF CONCRETE

1. The extra mass (greater inertia) and cracking (more damping) means the dynamic behaviour of concrete structures is better.
2. The extra mass is also an advantage as it reduces the holding-down requirement.
3. The stiffness is better especially if high strength concrete is used (as modulus and tensile strength are higher).

4. Better sound absorption and fire/impact resistance.
5. Finishing easier, e.g. flat plate apartments.
6. Higher thermal mass may lead to lower running costs.

3.9 DISADVANTAGES OF CONCRETE

1. Bigger requirement for lateral stiffness to avoid large P-delta effects.
2. Heavier solution so foundation costs will increase.
3. Integrating services with the structure may be difficult.
4. High strength concrete must be used to reduce column size.
5. Often not as fast to erect as steel (but if lead time for steel and time to enclosure is taken in to consideration the picture improves).

3.10 SUMMARY: CONCRETE SHEAR WALL BUILDING

1. Choose reasonable aspect ratio (e.g., about 5-7);
2. Choose gravity load system;
3. Use Eurocode 2 formula to estimate extent of shear walls;
4. Locate shear walls on plan to minimize torsions and ensure net stresses are compression;
5. Check second order effects due to lateral load;
6. Use code to estimate wind load and check P-delta effect due to lateral loads using MF formula;
7. Check accelerations by method in Eurocode 1 part 4 (but rarely a problem with concrete buildings.)

Note: if using other systems are used, e.g. tube, the drift implied by the Eurocode 2 formula must be not be exceeded, as the requirements for overall buckling is the same.

CHAPTER 4: STEEL AND COMPOSITE BUILDINGS

4.1 GENERAL

The floor is likely to be about 130 mm light-weight concrete built on trough decking. The minimum thickness is usually governed by the fire rating (often two hours for high-rise). Beams are usually designed to act compositely with the floor using welded studs.

Composite buildings attempt to use the advantages of steel and of concrete. In the tallest building in S.E. Asia, Petronas Towers, a steel floor is used. Perimeter beams are concrete. The steel beams are connected to the concrete structure using plates anchored into the concrete by studs.

Typical values of building density for preliminary design: (Schueller, 1986)

- For steel building 190-200 kg/m^3.
- For composite building 300 kg/m^3.

Typical values of steel weight: (Wells, 2005)

- World Trade Centre, New York: 244 kg/m^2
- Sears Tower, Chicago: 188 kg/m^2
- John Hancock Centre, Chicago: 148 kg/m^2

4.1.1 Floor Layouts

The layout may mean the primary beam required is deep compared to the secondary beam. (Figure 4.1).

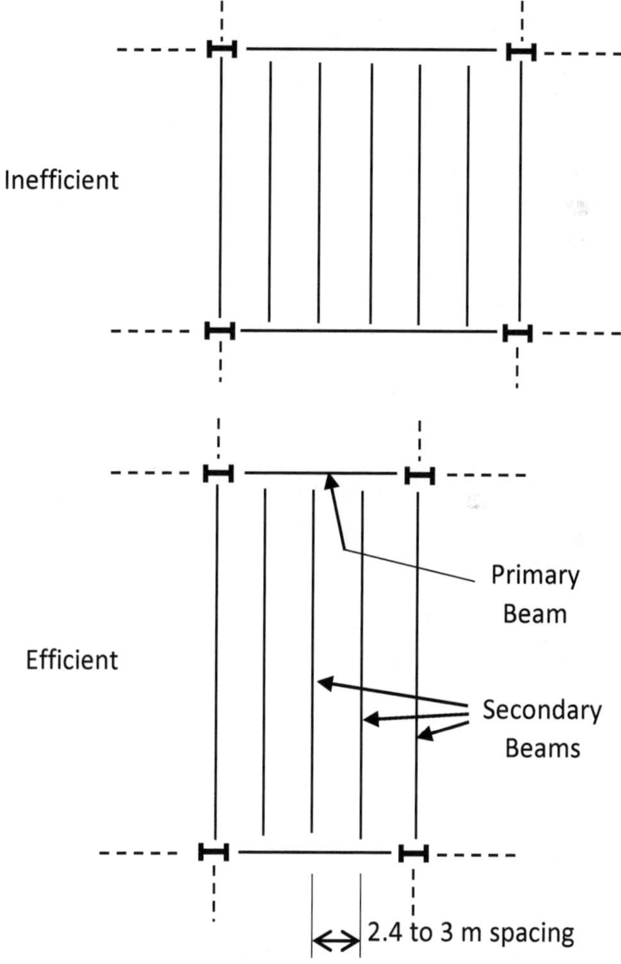

Figure 4.1: Rectangular layouts are to be preferred

Generally, the maximum span of steel beams is kept to about 10 m for economy. Camber beam if DL deflection exceeds 20 mm. Usually design so that LL deflection is less than span/360.

In an office building, there is usually a central core and a clear span from core to perimeter. Any internal columns are placed in the core area. This is to keep the perimeter-to-core space column-free. This span may be 13 m or more. Frequently, long span trusses rather than I-beams are used for this span. Such trusses may consist of T-sections as chords and double angles as web members. No gusset plates are needed if these sections are used. (Figure 4.2) These trusses can be used for spans up to 20 m. (Taranath, 1998)

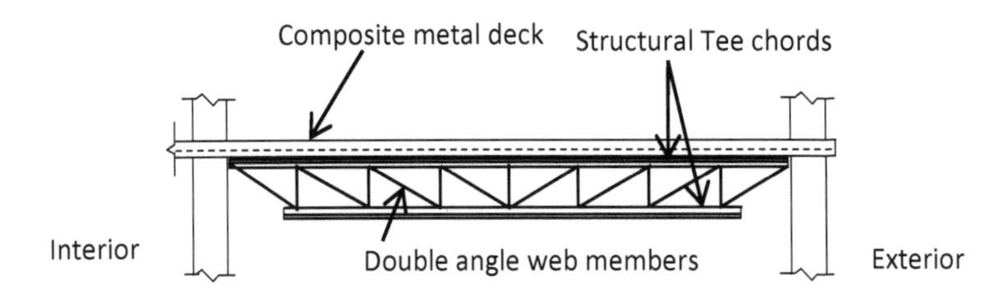

Figure 4.2: Long span steel truss

Care should be taken to ensure the natural frequency of such a long span truss is high enough to avoid vibration problems (certainly above 4 hertz).

4.1.2 Braced Frame

Bracing creates a vertical truss that behaves analogous to the shear wall considered in the previous chapter. Many forms of bracing are possible but the figure below shows cross bracing which is perhaps the most common. (Figure 4.3). Connections are simple as only shear is to be transferred.

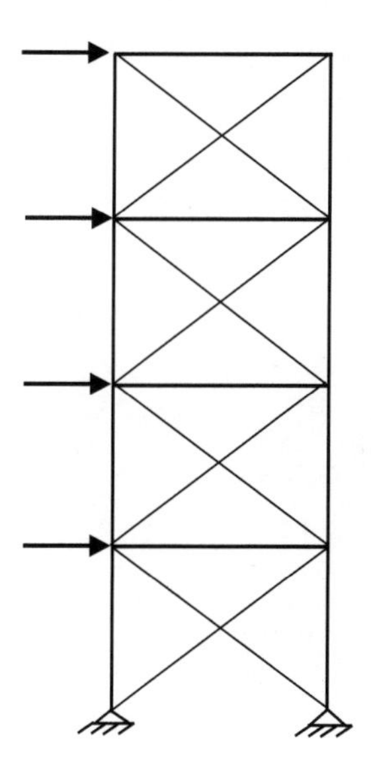

Figure 4.3: Diagonally Braced Frame

4.1.3 Properties of Braced Frame

- Area = $2A_c$
- Equivalent second moment of area, $I = A_c b^2/2$

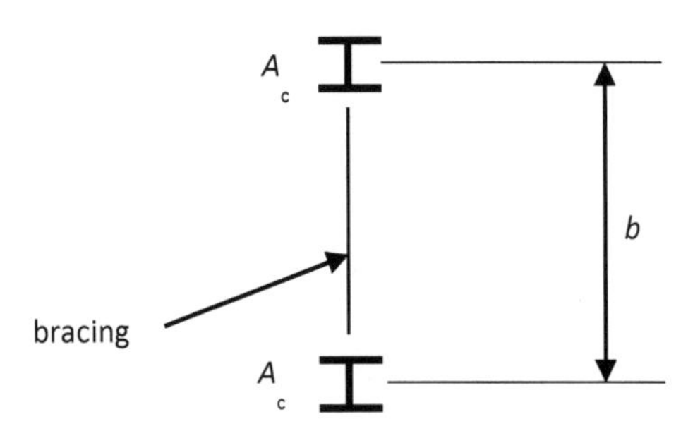

As with shear walls use a symmetrical layout if possible with a minimum of three braced bays (two in one direction). Typically the height-to-width ratio is less than about eight for maximum efficiency.

The calculating deflections of a braced truss are complicated as the diagonal members deform, thus the "shear" deflections are large (i.e., deflections due to the deformation of the web members are large).

The Interaction with concrete cores relatively complex.

If the building is less than about 40 storeys or so, it is common to use standard sections as columns. Otherwise built-up section using plates are used.

4.1.4 Advantages/Disadvantages of steel

- Speed of erection of the frame (although lead-time may mean date of weather proofing comparable to concrete solution).
- Easy to accommodate future modifications.
- No creep or shrinkage.
- Lighter-weight an advantage when considering foundations and magnitude of seismic load but a disadvantage considering holding-down and dynamic response.

4.2 COMPOSITE BUILDINGS

Many of the tallest buildings in the world are composite. They often have a similar plan, e.g. Jin Mao Tower, IFC 2, Taipei 101, Kowloon Station Tower (Figure 4.4):

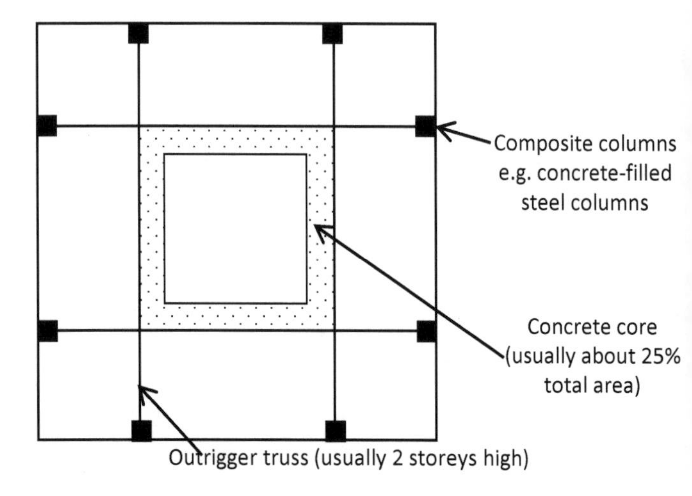

Figure 4.4: Typical plan of composite high-rise

4.2.1. Composite columns

Steel sections can be imbedded in concrete or steel tubes can act as permanent formwork for concrete. (See Figure 4.5).

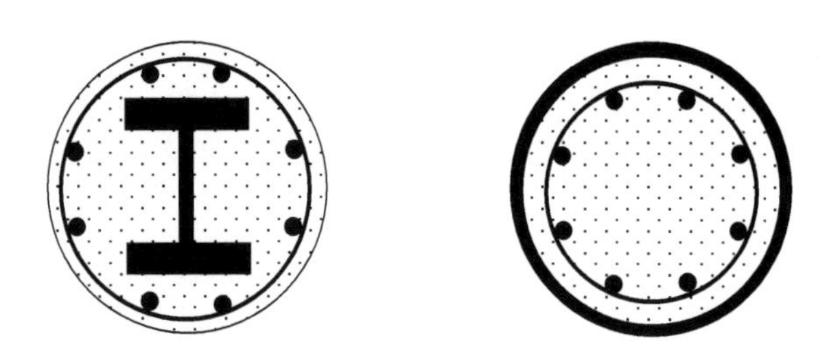

Figure 4.5: Typical composite columns

A concrete or composite column is usually much cheaper than all-steel one. The composite frame can be quick to erect if the steel is made the "erection frame" (i.e., take all loads during construction). The concrete follows later.

4.2.2 Second order effects

In Chapter 3 a formula, suggested by Eurocode 2 for the quantity of bracing elements necessary in a concrete building, can be used for composite buildings without modification, as their bracing is usually concrete. If the building is all-steel, the formula can, in principle, be used although modified (to include shear deflection).

4.2.3 After 9-11

(The Institution of Structural Engineers, 2002.)

It is recommended that:

1. More concrete is used in high-rise construction, e.g., most tall buildings since 9-11 use concrete core walls rather than the thin sheet-rock (gypsum board) used to surround stairs in the World Trade Centre. Concrete has better impact resistance.

2. Rely more on *passive* fire protection and ensure it is robust.

3 Provide pressurized stair wells.

4 Provide escape routes adequate in number and size, and as far apart as possible.

CHAPTER 5: CASE STUDY:

JOHN HANCOCK TOWER

- To illustrate the problems a poor preliminary design can cause.
- Engineers were caught-out by the change in building construction practice: use of less dead-weight and stiffness in a tall buildings.
- Thus wind now a much more important design criterion than previously.

John Hancock Tower, Boston:

(Cambell, 1995 & Levy and Salvadori, 1992).

Construction on the sixty-storey steel tube-in-tube structure (234 m) was completed in 1972. The architect was Henry Cobb (from IM Pei's firm). In 1974 the office received an award from the American Institute of Architects. A plan of the tower is shown in Figure 5.1.

The tower was completely covered in reflective double glazed glass, chosen to "hide" building's bulk. The glass panels were 1.35x3.45m. Glass panes of this size were never used on such a tall building before.

The tower satisfied all code requirements. One important requirement of high-rise construction usually not regulated by code is the limitation of wind-induced drift. The original engineer, who was an experienced high-rise designer, proportioned steel frame to limit deflection at top to around $H/400$ under the 50-year wind.

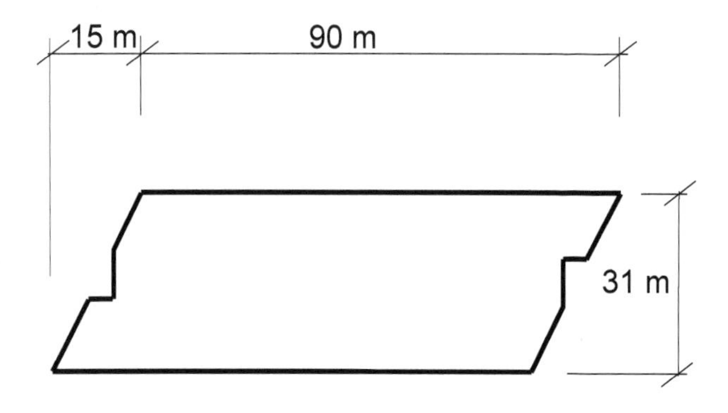

Figure 5.1: Plan of tower (each level similar)

Even before construction was completed glass panels started to fall. The building was nicknamed "plywood palace" as fallen glass was replaced by plywood. The building stood empty for 4 years while its problems were sorted out. The cost of delays and design modifications **doubled** original construction cost.

The owner insisted all parties maintain silence about building's problems. Nevertheless, the full story was revealed by 1995.

It was obviously a slender tower, having an aspect ratio (height/width) of 7.5. Therefore, the most obvious possibility was that glass was breaking because building was moving too much.

Experts were called in from MIT to advise and measure the actual movements. However, the measurements carried out on the building did not explain the glass breakage. It was found that displacements were large: so large that it was suspected safety of tower itself was in question because of the P-delta phenomenon. The suspicion was that there was a P-delta problem in the short direction of the building. Also accelerations were large. This would affect comfort of occupants.

New wind-tunnel tests were ordered. They were done in the boundary layer wind tunnel at the University of Western Ontario, Canada. This series of wind-tunnel tests showed (1) that wind sway was **not** the reason for the glass panel failure; and (2) that upper floors would experience a high acceleration in the short direction of the building and in twisting (the torsional natural frequencies and bending natural frequencies were similar, so the motions reinforced each other.)

A consultant, William LeMessurier, was hired. He had recently invented a damping system called a "tuned mass damper" for the recently built Citicorp Building, New York. He was called in to design a damping system for Hancock.

For the building he designed two dampers. Each was a 3,000 kN block of lead resting on a film of oil and attached to the structure using springs. Pumps were activated to float block when sensors detected a certain degree of motion.

A TMD is based on the principle of inertia. A heavy block tends to remain stationary as the building moves beneath it. Springs are attached to the block. Therefore, when the building moves, the springs ensure a restoring force is applied counteracting the movement. The springs are tuned to have same period as building.

The dampers dealt with the "comfort" problem, but, as mentioned previously, there was still a suspected P-delta "safety" problem. A world expert on building dynamics, Bruno Thurlimann, was consulted. He analyzed the tower and considered it too flexible for safety in the long direction of the building. According to him the P-delta problem was in this direction only.

To remedy this problem he proposed adding stiffness. However this conclusion was based on the assumption that the blockwork internal partitions did not contribute to lateral stiffness. The measured deflection was only $^{1}/_{3}$ of that based on this stiffness. The crisis of public confidence meant owner was extra cautious, and decided not to rely on this extra stiffness (as it probably was only available in the short-term). A total of 1650 tons of new steel diagonal braces were added to the frame's core to *double* stiffness in long direction of frame.

During all this glass continued to break. Eventually IM Pei's office discovered why. The reason was completely unrelated to the other problems of the building. Heating and cooling of glass panel itself was to blame. The glass panels were double glazed and the outer pane had a silver coating on the inside. The panes were separated all round by a continuous lead spacer. Thousands of cycles of heating and cooling caused a fatigue crack in this weld of this lead spacer. The bond obtained between lead and metal reflective coating was strong, so cracks migrated into the glass. Subsequent research by the architect revealed that reflective panels had given similar trouble on smaller buildings too. <u>All</u> of 10,344 identical panels were replaced with single thickness tempered glass.

Summary:

1. Too much acceleration in the short direction;

2. Too much twisting acceleration.

   These problems could be solved with TMD.

3. P-delta problem in long direction.

   This problem could only be solved by adding more stiffness in the long direction.

4. Glass breakage.

   Problem was with glass unit. Solution was to replace all the glass units.

5.1 LESSONS:

- Avoid multi-storey structures that are too flexible laterally.
- For all high-rise structures the P-delta check is very important.
- Simple limits on top deflection of multi storey buildings should be used with caution.
- Remember too that the P-delta effect is usually worst in the long-direction of the building.

- Unusual plan shapes can cause unexpected wind effects (e.g., torsion).
- Avoid similarity in the torsional and bending natural frequencies. This makes accelerations at the ends of the building especially noticeable. Building occupants tend to be particularly sensitive to torsional accelerations.
- High-rise buildings of more than about 120 m tall should be tested in a wind tunnel especially if the slenderness is high (> 5-7).
- These tests often pay for themselves in the long run as the overall forces may be less than code values.

5.2 ADDITIONAL INFORMATION:

- Early skyscrapers had masonry external walls and partitions, so were heavy and stiff.
- Thus, accurate determination of wind effects was of little importance.

  E.g. Empire state building: (mass 370 kg/m^3) about 20% of its stiffness comes from the steel frame and about 80% from heavy cladding and masonry partitions inside and the concrete encasement of the columns. (Robertson, 2005)

- The limit of Height/500 for top deflection was used to proportion the frame. Thus this limit dates from 1930s. (Robertson, 2005). Its use is perhaps not appropriate for modern high rise structures especially those in low wind environments such as Singapore as it leads to high Modification Factors.

5.3 Check P-delta effect due to lateral load in the long direction of the Hancock building

Modification Factor (MF) = 1/(1-*WR/Q*)

Apply formula to Hancock's data:

- Building height = 234 m.
- Building weight, *W* = 190x9.81x31x90x234 = 1217x10^6 N (assuming density 190 kg/m^3),
- Drift ratio, *R* = 1/400.
- If wind pressure is 1,400 N/m^2 on narrow face of building (31 m), Thus *Q* = 1400x31x234 = 10.1x10^6 N.
- Applying formula, Modification factor = 1/(1-1217/400x10.1) = 1.43

- Thus overturning moments due to applied lateral loads are 43% larger than would be calculated if P-Delta were neglected.

CHAPTER 6: FURTHER EXAMPLES

Q1: Costs:

Which is the largest single component of the construction costs of the RC frame of a typical building?

(1) Concrete,

(2) Steel reinforcing, or

(3) Formwork

Solution

(3) Formwork (usually costs 50%-60%)

Q2: Second order effects (1):

The frame shown below deflects under the load *P*.

Write expressions for the first order base moment, and the approximate value of the total base moment.

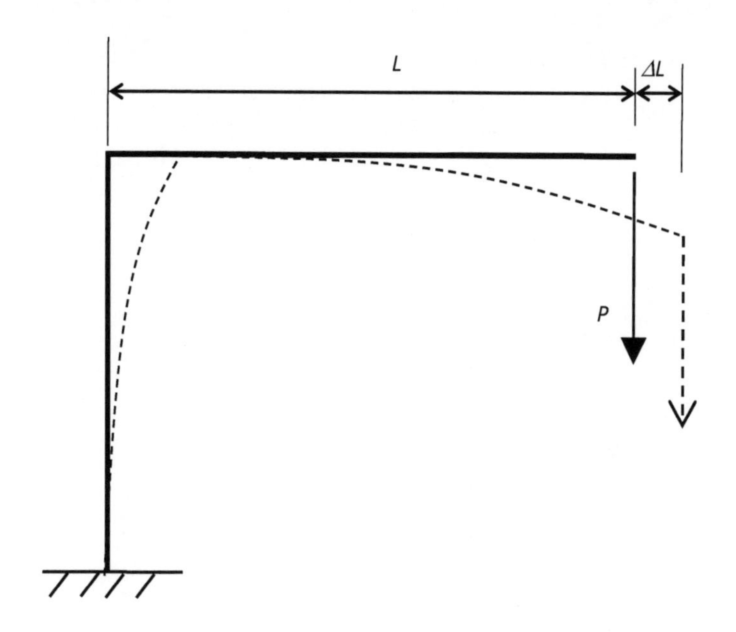

Solution:

First order base moment $M_1 = P\,L$

Including only 2nd order effects due to large deflections:

$M_{base1} = P\,(L + \Delta L) = M_1\,(MF)$

Including only 2nd order effects due to gravity load (neglecting self-weight):

$M_{base2} = M_1\,/\,(1 - P/P_{cri}t)$

M_{TOT} = max [M_{base1}, M_{base2}]

Q3: Second order effects (2)

Consider this hypothetical 30-storey concrete condominium building built in Singapore/Malaysia. The lateral frame is designed using a drift ratio *H*/500. Compare the MF in both directions.

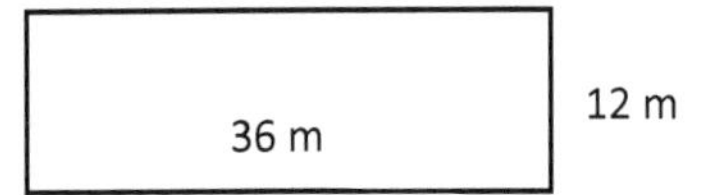

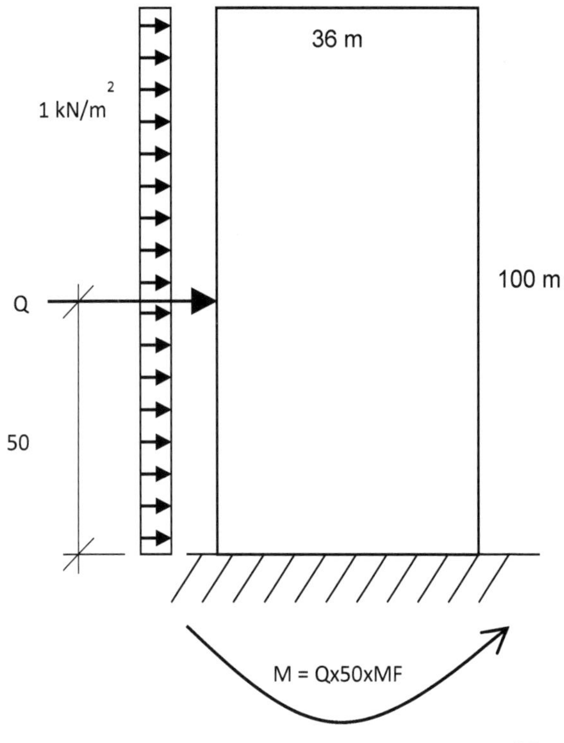

Solution:

Wind on narrow face, i.e., 12 m; Q = 1000x12x100 = $1.2\text{x}10^6$ N

Building Density = 400 kg/m^3; W = 400x12x36x100x9.81 = $1.7\text{x}10^8$ N

R = 1/500

$MF = 1 / (1\text{-}W.R/Q)$

= 1/ (1- $1.7\text{x}10^8$/500x$1.2\text{x}10^6$) = 1.39

Wind on broad face, i.e., 36 m; Q = 1000x36x100 = $3.6\text{x}10^6$ N

Building Density = 400 kg/m^3; W = 400x12x36x100x9.81

= $1.7\text{x}10^8$ N

R = 1/500

$MF = 1 / (1\text{-}W.R/Q)$

= 1/ (1- $1.7\text{x}10^8$/500x$3.6\text{x}10^6$) = 1.10

Q4: Second order effects (3)

The following 30-storey concrete building is designed using drift ratio H/500.

Compare the values of MF calculated in the long direction, if this building is designed to resist a wind load of (a) 800 N/m^2 or (b) 2500 N/m^2.

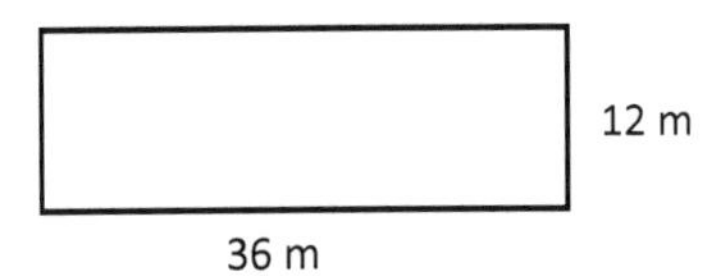

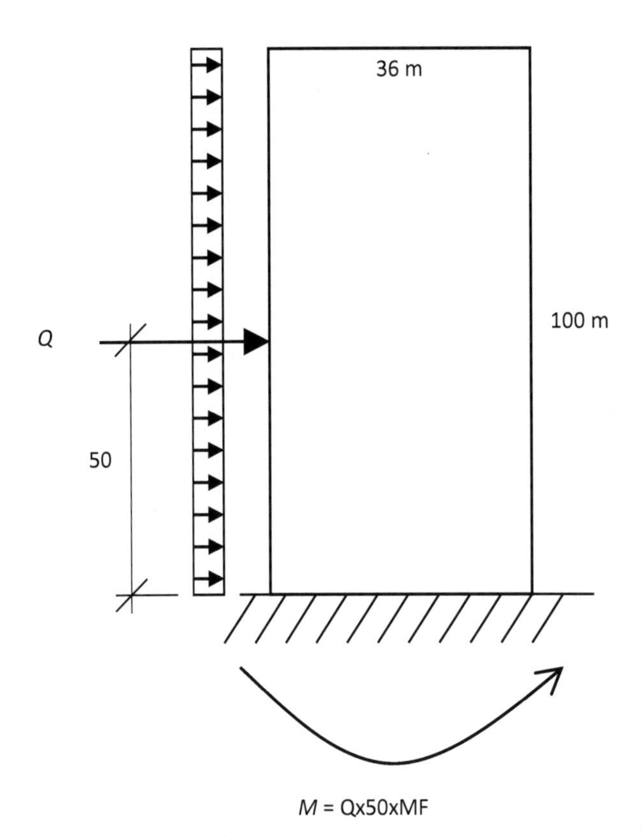

Solution (a)

Wind on narrow face:

Q = 800x12x100 = 9.6x10^5 N

Building Density = 400 kg/m^3 (typical concrete building)

W = 400x12x36x100x9.81

= 1.7x10^8 N

MF = 1 / (1-WR/Q)

= 1/ (1-1.7x10^8/500x9.6x10^5) = 1.55

Solution (b)

Wind on narrow face:

Q = 2500x12x100 = 3x10^6 N

Building Density = 400 kg/m^3 (typical concrete building)

W = 400x12x36x100x9.81

= 1.7x10^8 N

MF = 1 / (1-WR/Q)

= 1/ (1-1.7x10^8/500x3x10^6) = 1.13

Q5 (a): Shear Walls

Assuming the diaphragm is rigid, what proportion of load shown is transferred to each shear wall?

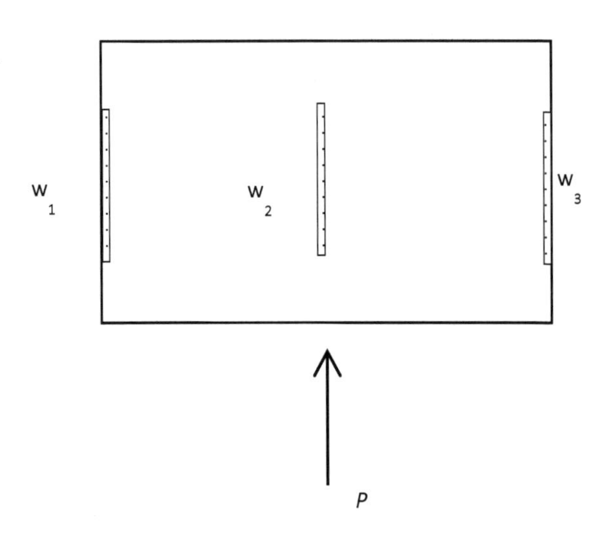

Solution

- W_1 - 33%
- W_2 - 33%
- W_3 - 33%

Q5 (b): Shear Walls

Assuming the diaphragm is rigid, what load is resisted by each shear wall? All walls 200 mm thick.

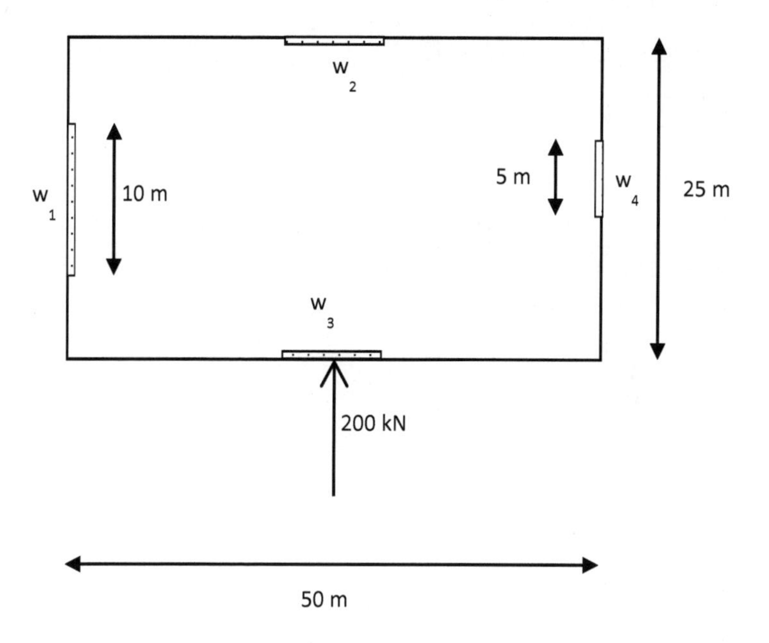

Solution

Layout of walls is unsymmetrical, so find position of shear centre:

$$y = \Sigma I.y/\Sigma I$$

$$= (0.1\text{x}0.2\text{x}10^3/12 + 49.9\text{x}0.2\text{x}5^3/12)/(0.2\text{x}10^3/12 + 0.2\text{x}5^3/12)$$

$$= (1.67+104)/(18.75) = 5.64 \text{ m}$$

Torsional moment = 200x(25-5.64) = 3,872 kNm

Load to w_3 (w_2 similar) = 3872/24.8 = 156 kN (assume all of the torsion is resisted by these walls).

- w_1 - 100 kN
- w_4 - 100 kN
- w_3 - 156 kN
- w_2 - 156 kN

Q6: Hancock

Compare value of P-delta effect on narrow face calculated using formula (1.43) and that which is obtained by considering all the mass at *H*/2

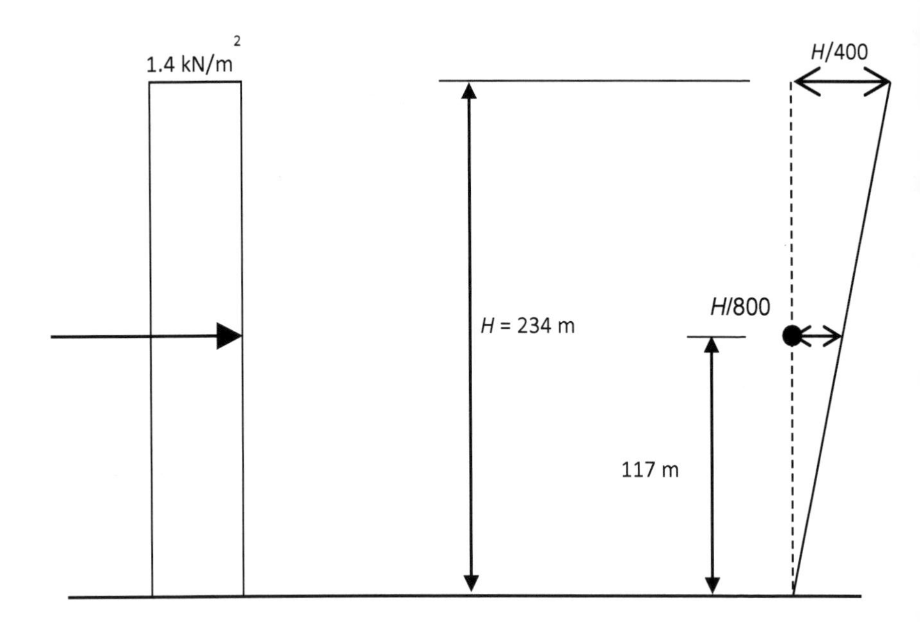

Solution

Moment at base due to wind = $1.4 \text{x} 31 \text{x} 234^2/2 = 1.19 \text{x} 10^6$ kNm

Weight of building = $1217 \text{x} 10^3$ kN

Considering all the mass concentrated at H/2, thus additional moment at base = $1217 \text{x} 10^3 \text{x} 234/800$ = $356 \text{x} 10^3$ kNm

Thus total moment at base = $1.54 \text{x} 10^6$ kNm

Total/wind = 1.54/1.19 = 1.29 (11% difference)

Note: Taking mass *M*/3 at *H*, *M*/3 at 2*H*/3 and *M*/3 at *H*/3 gives MF = 1.4

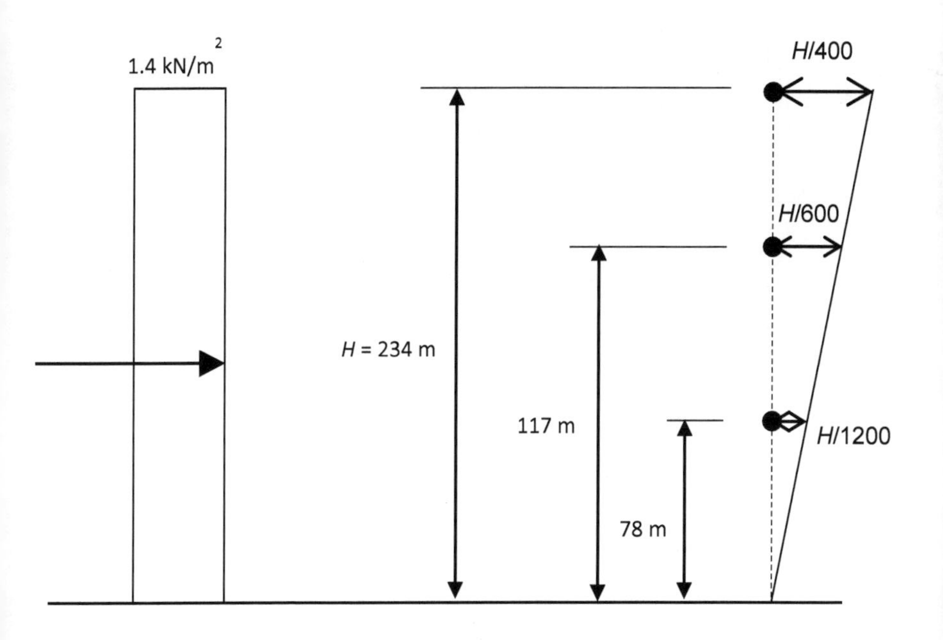

M_{add} = 475 x10^3 kNm

Total/wind = 1.67/1.19 = 1.4

(2% diff)

Q7: Wind

For the building below where is the maximum wind speed, and so pressure, likely to be experienced?

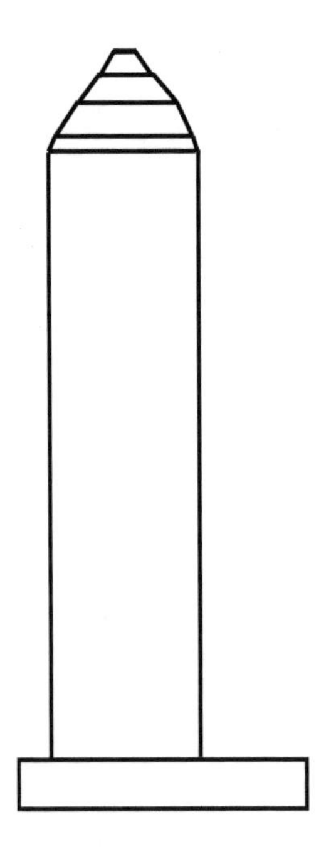

Solution:

(Tomasetti et al, 1987)

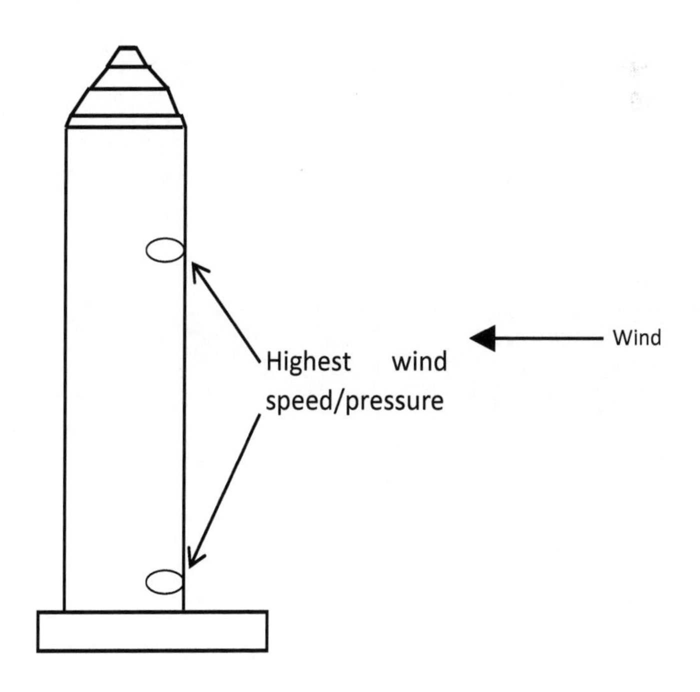

Q8: Second order effects (4):

A concrete building is 40 m x 40 m on plan. If it is 30-storeys high (100 m), what shear walls are necessary according to Eurocode 2? Assume concrete f_{cu} = 45

N/mm^2. Are these walls sufficient to control lateral load P-delta effects if the 50 year wind is 1 kN/m^2? Assume the walls are uncracked. Take E_{cm} = 34 kN/m^2. Assume purely flexural behavior (i.e. shear deflections are negligible).

Solution:

$\sum E_{cm}I_c \geq F_v(n + 1.6)\, H^2/0.517n$

F_v = 400x40x40x100x1.5 = 941x10^6 N

$\sum E_{cm}I_c \geq$ 941x10^{6}x31.6x100^2/(0.517x31) = 1.86x10^{13} kNm^2

E_{cm} = 34x10^6 so I_c = 545 m^4

Try 14x14 core using 0.4 m thick wall

$I = \frac{1}{12}(16^4 - 15.2^4)$ = 671 m^4 > 545 m^4 (extra useful for allowing for holes) => okay.

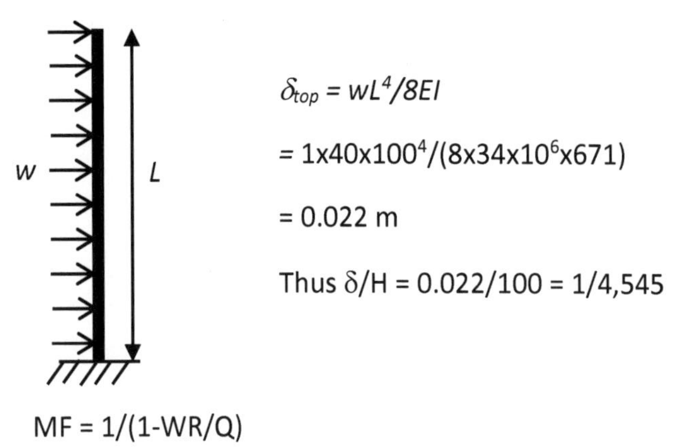

MF = 1/(1-WR/Q)

$W = 400x9.81x40x40x100 = 628x10^6$ N

R = 1/4,454

$Q = 1x40x100x1000 = 4x10^6$ N

Thus MF = 1/(1-628/4545x4) = 1.04 which is less than about 10% and so acceptable.

Q9 Walls

For the data of the previous example calculate the likely value of the service stresses in the walls when the 50-year wind is blowing. Take the service load as 10 kN/m^2.

Solution

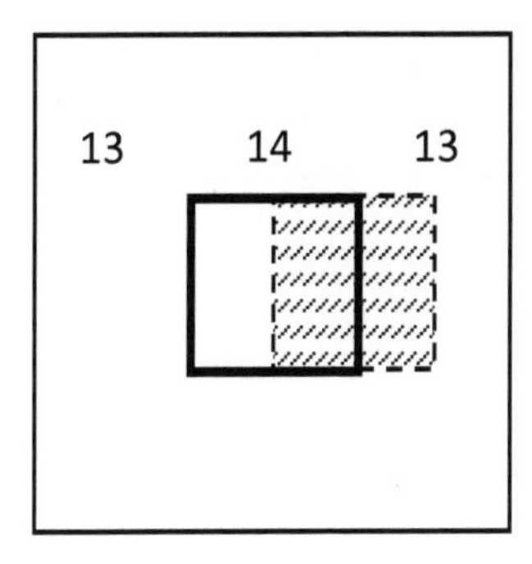

Tributary width= 13.5 m

Consider 1m width of wall. Service load = floors + s/wt wall = 30x10x13.5 + 0.4x25x100 = 4,050 + 1,000 = 5,050 kN,

Thus $\sigma_v = 5,050x10^3/400x1000 = 12.6$ N/mm^2

Wind moment, $M_w = 40x100^2/2 = 200,000$ kNm

Stress, $\sigma_w = M_w y/I = 200x10^9x7,000/671x10^{12} = 2.1$ N/mm^2.

Thus net stress is likely to be a compression of about 10.5 N/mm^2 i.e. likely walls are uncracked.

Q10: Second order effects (5):

Assume the same data as in Q8 except the building plan's aspect ratio is 2:1 instead of 1:1, i.e. plan is 20m x 80m. How does this affect the provision of walls?

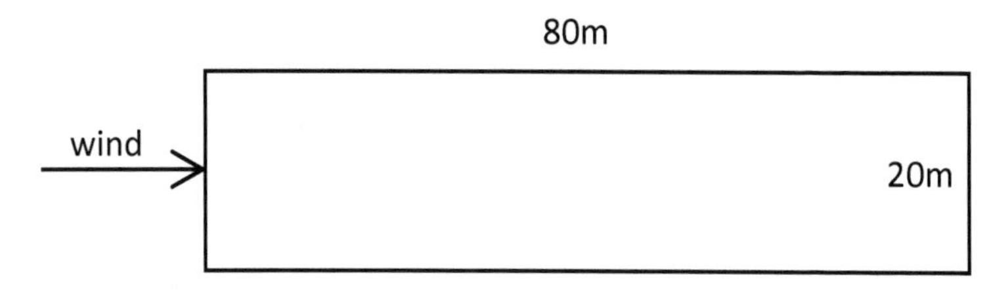

F_v is unchanged so I_c required is unchanged. Thus δ_{top} is unchanged.

The value of MF in the long direction of the building is:

1/(1-WR/Q)

W = 628×10^6 N

R = 1/4,454

Q = 1x20x100x1000 = 2×10^6 N

1/(1-WR/Q) = 1/(1-628/4,454x2) = 1/(1-0.07) = 1.08 < 1.1 => ok

Thus no change is necessary.

Note: if aspect ratio of plan is 1:6, then MF = 1.1. For greater aspect ratios, more walls need to be provided in the long direction.

Appendix:

Table A1: Deflection formulae for cantilevers: (McLeod, 2005).

System (W is total load)	Bending Moment Diagram	Maximum Bending Deflection	Maximum Shear Deflection
W L	$WL/2$	$WL^3/8EI$	$WL/2A_sG$
W L	$2WL/3$	$11WL^3/60EI$	$2WL/3A_sG$

Note: G = E/(1+ν) where ν is Poisson's ratio (0.15 for concrete and 0.3 for steel).

And A_s for a rectangle of area A is $5A/6$.

References

American Concrete Institute, ACI-318-14, *Building Code Requirements for Structural Concrete*, 2014.

Billington, D, Goldsmith, M., *Technique and Aesthetics in the Design of Tall Buildings*, Fazlur Kahn Memorial Session, Institute for the design of the high-rise urban habitat, Lehigh University, 1986.

Bourne, S., *Prestressing: recovery of the lost art*, The Structural Engineer, Vol. 91, February 2013.

British Standards Institution, BS EN 1991:2004, *Eurocode 1: Actions on structures*, 10 parts, 2005.

British Standards Institution BS EN 1992-1-1:2004, *Eurocode 2: Design of concrete structures* — Part 1-1: General rules and rules for buildings, 2005.

Broker, O., *Concrete Buildings Scheme Design Manual*, 2006, The Concrete Centre, UK.

Cambell, R, *Builder Faced Bigger Crisis Than Falling Windows*, Boston Globe, 3 March 1995.

CIRIA, *Design of Shear Wall Buildings*, Report 102, 1984, UK.

CIRIA, *Design for movement in buildings*, Report 107, 1981, UK.

Dowling, P. et al., *Structural Steel Design*, Butterworths, 1988.

Eisele, J., and Kroft, E., *High-Rise Manual: typology and design, construction and technology,* Birkhauser, Switzerland, 2003

Fling, R., *Practical Design of Reinforced Concrete,* Wiley, 1987, USA.

Grossman, J., *Slender Concrete Structures-the new edge,* ACI Structural Journal, Vol. 87, No. 1, January-February 1990, pp. 39-52.

Heyman, J., *Basic Structural Theory,* Cambridge University Press, Cambridge, UK, 2008.

Levy, M. & Salvadori, M., *Why Buildings Fall Down,* Norton, 1992.

MacLeod, I., *Modern Structural Analysis: Modelling Process and Guidance*, Thomas Telford, 2005.

Millais, M., *Building Structures*, E&F Spon, UK&US, 1997.

Robertson, L., See, S., *Preliminary Design of High-Rise Buildings,* Building Structural Design Handbook, Ed. White & Salmon, Wiley, NY, 1987.

Robertson, L., *Rising to the challenge*: IStructE Gold Medal Address, Presented at a lecture and dinner held at The East Winter Garden, Canary Wharf, London on 9 March 2005.

Schueller, W., *High-Rise Building Structures*, 2nd edition, Robert E Krieger Publishing, Florida, 1986.

Simu, E. & Scanlan, R., *Wind Effects on Structures*, 3rd edition, Wiley, 1996.

Stafford-Smith, B., Coull, A., *Tall Building Structures: Analysis* and Design, Wiley, 1991

Taranath, B., *Steel, concrete and composite design of tall buildings*, 2nd edition, McGraw-Hill, 1998.

The Concrete Society, *Post-tensioned concrete floors: design handbook*, Technical Report TR43, Second Edition, 2005, UK.

The Institution of Structural Engineers, *Safety in tall buildings and other buildings with large occupancy,* The Institution of Structural Engineers, London, July 2002.

Timoshenko, S., *Theory of elastic stability*, 2nd edition, McGraw-Hill, 1961.

Tomasetti, R., et al, *Tall Buildings-Load effects and special design considerations,* Building Structural Design Handbook, Ed. White & Salmon, Wiley, NY, 1987.

Wells, M., *Skyscraper: structure and design*, Yale University Press, 2005, USA.

About the author:

Er. Dr. Niall MacAlevey is currently an independent consultant specializing in the analysis and design of reinforced and prestressed concrete structures, forensic engineering and the strengthening of concrete structures. He is the founder of the firm "Shamrock Consultants", and is a registered Professional Engineer in Singapore. He graduated from University College Dublin, Ireland in 1987, and completed his M.Sc. degree in "Concrete Structures" at Imperial College, London. He completed his Ph.D degree at the Nanyang Technological University in 1997 on "The Strengthening of Concrete Structures" and later joined the academic staff there. He obtained a PGDipTHE (Post-Graduate Diploma in Teaching in Higher Education) from the National Institute of Education in 2001. He has worked for a number of consulting engineering firms and specialist prestressing subcontractors in London, Cambridge, Hong Kong and Singapore.

He can be contacted at niallmacalevey@gmail.com

Printed in Great Britain
by Amazon